D0847675

Honey in the Hive

Etta Nommensen

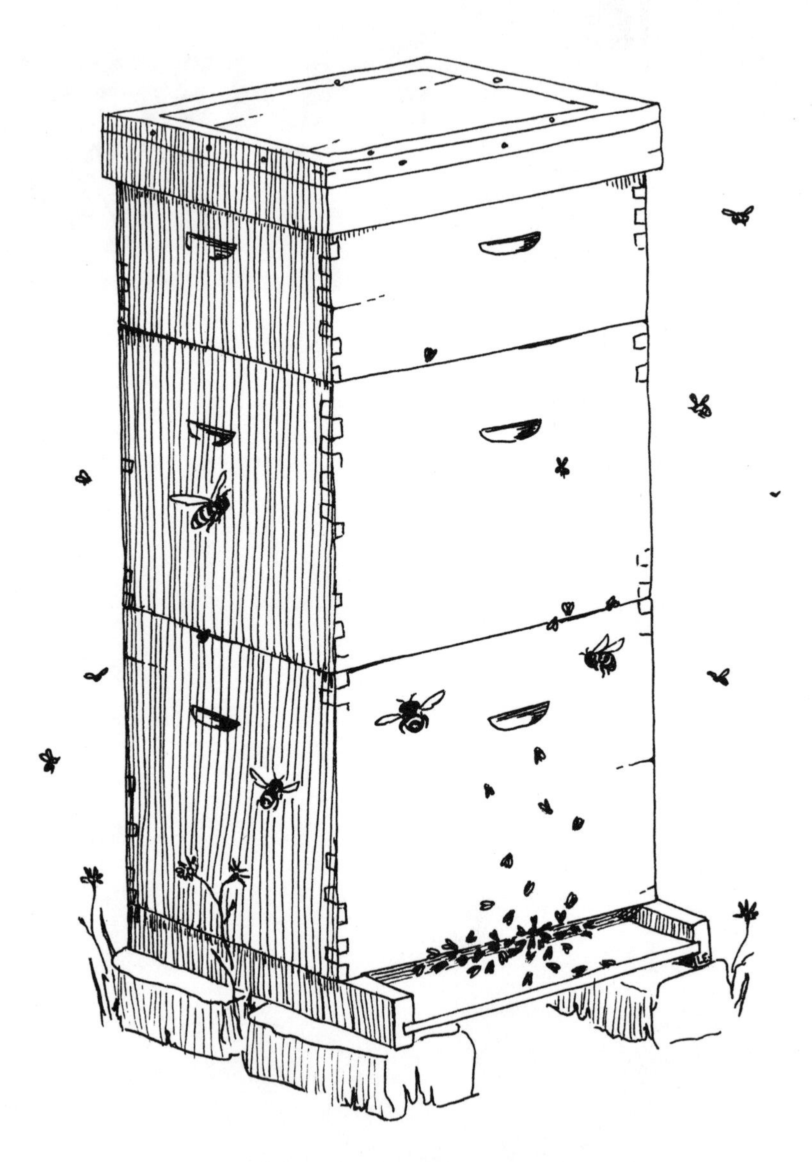

Modified Dadent hive with double brood chamber.

Honey in the Hive

A Beekeeper's Journal and Guide

Etta Nommensen

Illustrations by Douglas Stuart Wells and Lila Eisberg

WINCHESTER PRESS

Library of Congress Cataloging in Publication Data

Nommensen, Etta Douglas.

Honey in the Hive

Includes index.
1. Bee culture. I. Title.
SF523.N757 638'.1 79-13406
ISBN 0-87691-308-7

Published by Winchester Press
1421 South Sheridan
Tulsa, Oklahoma 74114

Book Design by Marcy Schusterman

Printed in the United States of America

1 2 3 4 5 84 83 82 81 80

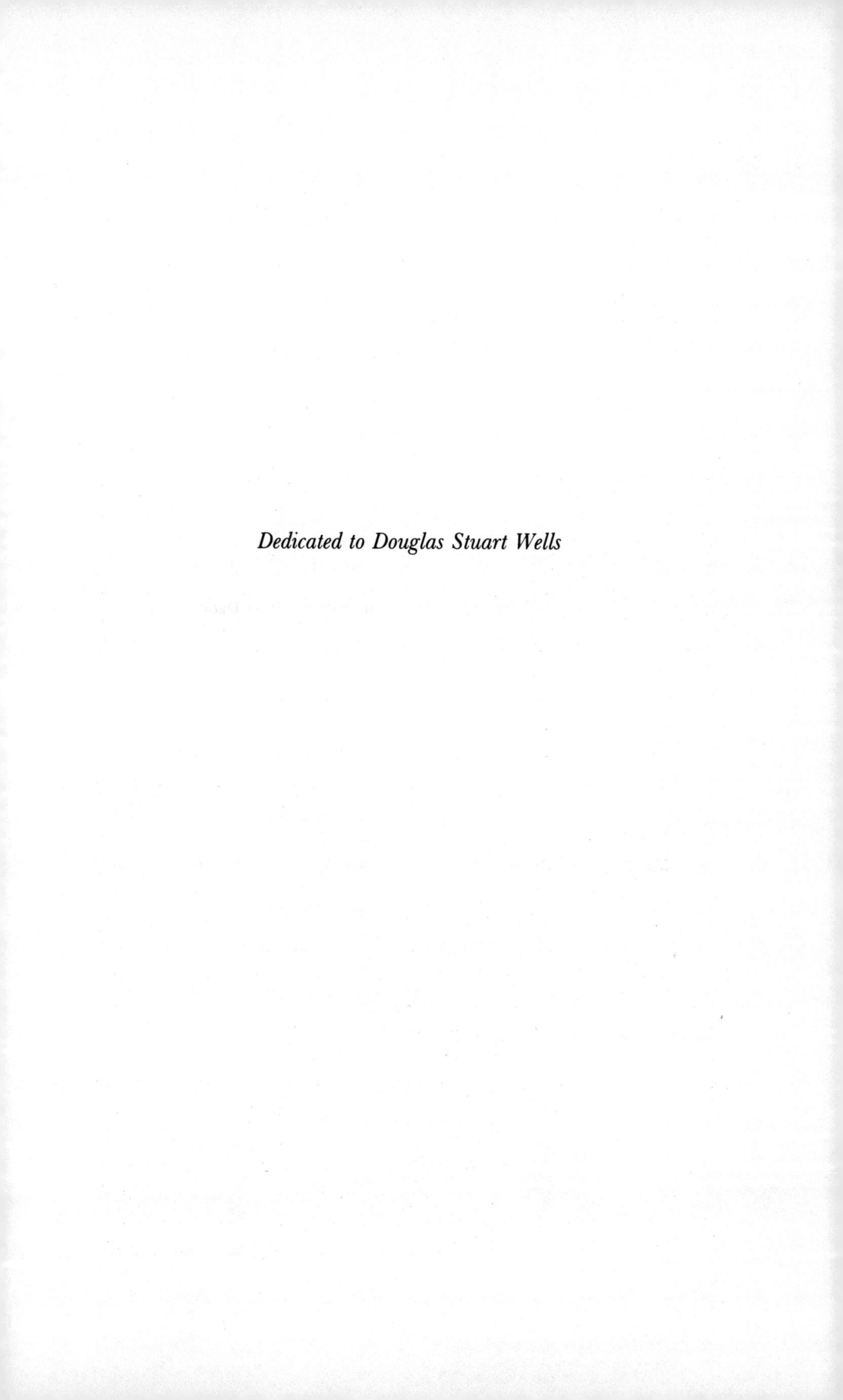

Dedicated to Douglas Stuart Wells

Contents

Foreword

I have been a beekeeper for fourteen years and my experiences have been interesting and varied. The first eleven years were shared with my partner Dave, and they brought me utmost joy. He was an excellent partner and I was sorry to lose him when he left for college. I'm sure he enjoyed the work as much as I did.

I find the cleverness and exactness of such small insects, as set forth in this work, most amazing. Bees are our most valuable insect, for in addition to furnishing us with honey, they pollinate the fruit trees, crops, and flowers. They are an important part of our ecology, yet few people know much about them.

Therefore, in this work I have endeavored to set forth little-known and interesting facts about them as well as relate my personal experiences with them. I hope to dispel the erroneous impression, held by many, that bees are vicious and will attack anyone who comes near their hive.

However, I realize that no one but a beekeeper would delve

into a textbook for information, so I have woven the facts into an account of our work with bees.

My information is taken largely from Mr. Langstroth's book *The Honey Bee*, as revised by Dadant. Not much has changed in the handling of bees since this book was first published in 1919.

I Meet the Bees

We lived on a wind-swept farm on the rolling plains of Colorado east of Denver, known as Windsor Heights, until I was six years old. Mother kept bees in a very simple fashion, and my first recollection of them was a day in May when I was four. Mother was prepared to take a swarm that was leaving a hive.

She wore a pair of dad's overalls, tucked into her high-buttoned shoes, and a flannel shirt with the cuffs stuffed into leather gloves. On her head, for a bee veil, was a strange contraption made of a corn sifter and a pillowcase. I had watched as she laboriously made this one evening.

The corn sifter was about twelve inches in diameter and four inches deep. The sides were of lightweight plywood, formed into a circle and stapled together where the wood was lapped over. The bottom was of screen wire, and many times I had used it to sift the fine meal from cracked corn so we would have cornmeal for bread or for food for the baby chicks.

In the hemmed end of the pillowcase mother put a draw string, to be tied around her neck as tightly as comfort would allow. The other end she ripped open and stitched to the rim of the corn sifter. She would look through the screen wire and the pillowcase kept the bees out of her hair.

So there was mother, dressed in this garb, beating a tin pan with a spoon and throwing sand in the air among the bees to make them light somewhere so she could put them in a hive. And light they did, but not before my father, who had been called to beat a pan and throw sand, had been stung on the nose. I remember that for several days he had the largest, reddest nose that I had ever seen. I was very much in awe of bees that could do that to my father.

Mother had made the hive from an apple box turned upside down on a one-by-twelve board. One end of the box had a three-eighth-inch strip cut from it where it met the bottom board, so the bees could go and come. She had also cut a three-sixteenth-inch slot about ten inches long in what was now the top of the hive, and put a smaller box upside down over it. The slot was to prevent the queen from entering the smaller box and laying eggs in our honey. The worker bees were small enough to go through this slot, but the queen wasn't. This upper box is now called a super and is the same dimension as the hive, but mother had the right idea because we had honey with no eggs, larvae, or baby bees in it. When the small box was full, mother would remove it and put another in its place without disturbing the bees.

Another vivid recollection of bees was a year later when a swarm made its home under our side porch, which cleared the ground by about three feet at one end as the house was built on the side of a small hill. Mother had not tried to hive the bees, since she had a couple of colonies in hives. That fall I was sitting on the front steps when a neighbor, Jed Anders, drove up in an old mule-drawn cart. He wore torn coveralls of faded blue and his red hair certainly needed combing.

He called to me, "Sis, is your mother handy?"

I ran up the steps and said, "Mom! Jed's here and he wants to see you."

Mother came to the door drying her hands on her apron as she said, "Good morning, Jed."

"Mornin', ma'am." he said. "Would you mind if I got the honey from the bees under your porch, seein' as how you got honey in your hives?"

"How could you do it?" mother asked. "I wouldn't want the porch floor torn up."

"No, ma'am!" said Jed, running a hand through his already tousled hair. "I wouldn't tear the floor up, I'd just smoke 'em out. I'd light some papers and rags in a bucket I got in the cart and let the smoke go under the porch."

I wondered if the bees would fly out carrying their honey, and he would put them in a box like mother did.

Mother hesitated, then said, "Jed, are you sure that you wouldn't set fire to the porch floor? I wouldn't want that to happen."

"Oh, no, ma'am!" Jed's red hair shone in the sun. "I've done it lots of times. I call it brimstoning. I got this big bucket in the cart and no flames come out, jist smoke. You see, the bottom of the bucket has holes in it that I punched for the draft, and I fixed an old funnel over a hole in the lid and the smoke just pours outen that when I light it up." Jed waited for mother to answer, but added, "We-uns sure need the honey, as we got no money to buy sugar."

"Well," mother said slowly, "if you're sure there won't be any flames I guess you can try, but do be careful. I'll bring a pail of water just in case something starts to burn."

Jed's face lit up as he said, "Oh, thank-ee, ma'am. I'll give you half the honey."

"We'll see what you get before we talk about it, but do be careful." Then mother turned to me, with a worried look on her face, and said, "You stay here on the steps and if anything happens, if even a tiny flame shows up, you call me at once and I'll pour the water on it. I must go back to the kitchen because the bread is ready to put into the pans."

Mother put the pail of water on the porch and I sat on the steps and watched Jed get ready to smoke the bees. He put the bucket by the open end of the porch, it was the biggest bucket I had ever seen and the bottom looked like a colander, it was so full of holes.

Jed tore some newspaper and put it into the bucket with an old red rag that looked like a piece of an old shirt. He put something else in too but I did not see what it was because he took it out of a pocket and slipped it in. But I did see that it was small and wrapped in newspaper and I thought it must be something to make a lot of smoke. He lighted the things in the bucket, turned it on its side, and put the lid on. In a minute, smoke poured out the funnel but there was no flame, so I didn't call mother. I felt very sorry for the bees, breathing all that smoke.

Jed pushed the bucket under the porch with a stick and I could hardly keep my seat on the steps, the smoke was so thick. But no bees came out. I got down from the steps and looked under the porch. The bucket was red like our heater sometimes gets in winter, but it was on dry sand so nothing caught fire. After the paper, rags, and that thing Jed had added had quit burning and the bucket cooled off, he rolled it away and kneeled down to look under the porch.

"That got 'em, I knowed it would," he said. He got an old washtub from the cart and crawled under the porch. I heard a lot of shuffling around and at last he crawled out, pulling the tub after him. He was dusty and sooty and when I saw the tub and what was in it, I screamed. Mother heard me and came running to the door thinking something was on fire, I guess, for she was white as a sheet.

Jed was all smiles, "See, I done it," he said, brushing sand out of his hair. "All that's different under there is a little soot and no bees. Now, if'n you'll bring a pan I'll give you some of the honey."

Mother looked at the tub in horror and managed to say, "No thank you, Jed. You keep it, we have plenty," and she went back into the house.

I was as horrified as mother as I looked at what was in the tub. I had never seen anything like it. There was a mass of partly melted wax, dead bees and some still struggling, larvae and eggs, all in the honey. I felt sick to my stomach and got up and went into the house, leaving Jed to load it into his cart and go away.

"Mother, why did you let him do that?" I asked. "Did you know he would kill all the bees like that?"

"I had an idea it would be like that," she said sadly.

"Then why did you let him do it?"

"He needed the honey, dear, and the bees would no doubt have died this winter when it got cold."

"I think it was a terrible thing to do, and I hate Jed and you too." I ran to my room and cried until I felt better. From that incident was born in me an intense love for bees, God's small insects that men killed so heartlessly for their honey.

In the summer of the year that I was six, we moved to the mountains west of Denver. Mother had received some money when granddad died and she bought 160 acres of land on Parmalee Gulch, which was about five miles above Morrison. We arrived after dark and in the morning when I looked out the window I called to mother, "Oh, come and see all the mountains we bought!"

She came and put her arm around me as she said, "Yes, dear, I know. Aren't they beautiful?"

Our house was a small log cabin with two towering spruce trees watching over it and there was an alfalfa field above it and below it on the sloping hill. It was an ideal place for bees and mother soon had two colonies which she took from a fence post or a stump where they had clustered as they came from a nearby cottonwood tree, on the bank of Parmalee Gulch, that had bees in it.

Its circumference was at least six feet and the bees were about thirty feet up in a fork of the tree. I would lie on my back in the shade and watch the bees, dark little spots drifting around against the green leaves. You had to know just where to look to see them.

Mother's two colonies were at the edge of the alfalfa fields and for a closeup view I would sit in the shade of a nearby tree and watch the bees leave for the alfalfa and return with their pollen sacks bulging with the yellow load. Other bees would meet them on the little porch and take their loads, or they would hurry inside. They were very busy yet never seemed tired with all the flying.

When I was nine we moved to Golden and mother gave her bees to a neighbor before we left the mountains. That was the last time mother kept bees; for she had none in Golden or Denver, where we moved after several years. There I finished my schooling and was married.

I did have a short brush with bees in my fifth-grade class early in May before we left Golden. The boy who sat across the aisle from me came into the room with a bulging pocket. As he slid into his seat he gave me a sly wink and opened his pocket to reveal a jar with bees in it. I clapped my hand over my mouth and felt my eyeballs pushing against their lids. He laid a finger on his lips and bent over a book.

Suddenly the girl in front of me screamed and began to slap

her hair and across my desk crawled a big drone, bigger than the one I saw in her hair. Then the room was in an uproar as bees were flying everywhere, even around the teacher. She was banging her desk with a ruler and shouting for order. As the bees found the windows and settled there, comparative calm seeped into the room.

Teacher demanded, "Who is responsible for bees in this room?" On getting no response she sailed up and down the aisles, probing pockets and peeking into desks, so it was inevitable that she should find the jar in Edgar's desk. And what was worse, it contained incriminating evidence—a lone bee that had not gotten out.

She took Edgar by the ear and propelled him to the front of the room, reviling him for exposing the class to the dangerous sting of bees. He was unsuccessfully trying to explain that there was no danger because they were drones and they have no sting. "Now don't lie besides all the commotion you've caused and endangered the class," she fairly shouted.

Edgar was cringing before her, insisting that there was no danger because drones can't sting. It was only a prank. She grabbed a ruler and flailed out at him, striking him on the head, shoulders, and even his face. The students were huddled in their seats, mute with fear and amazement.

It was then I arose to my feet and said in a voice I'd never heard before, "Miss Bliss! He's right! Drones can't sting and all the bees are drones. There was no danger, just excitement."

Miss Bliss stopped, ruler brandished for another stroke. "And just how would you know, young lady?"

"My mother always kept bees and I know about them." To prove it I dumped the drone out of the jar into my hand and took him up to her. I showed her his smooth round, stingless abdomen and said, "See, Miss Bliss, no sting."

"Both of you go to your seats," Miss Bliss said with no apology to Edgar. In a few minutes the class was back to normal, although a bit subdued. Edgar was wiping his bruised hands and face with his handkerchief, then he reached across the aisle and squeezed my extended hand in thanks for his deliverance.

I whispered, "No more bee jokes," and he slyly winked a swollen eye.

A Cub Scout and Bees

Some years after our marriage, my husband was transferred to Joplin, Missouri, by the company for which he worked, and we lived there for a number of years. When the two older children were through college and married, Dave, who was eight, became a cub scout and bees again came into my life.

Each scout was told to choose a hobby or project and Dave chose bees, but he came home from the meeting quite annoyed.

"When I chose bees, the den mother threw her hands up in horror and asked me, 'Why bees? The other boys chose gardening, wood burning, leather work, or nature study.'

"Then I asked her if bees weren't nature study, and she said I'd be on my own because she knew nothing about bees." He flung his cap in a corner in disgust.

But bees it was for David. What the den mother didn't know was that Dave had been a nature boy all of his life and our house had been home to little lizards, butterflies, fireflies, junebugs, and even praying mantises. The yard had white giant ducks that had arrived each Easter in the form of fuzzy balls on yellow legs. There was also a huge black rabbit that Dave had accepted when it was tiny because someone was going to kill it if he didn't keep it. And now Dave wanted bees.

At length the den mother arranged to take the boys to see a beehive as an educational project, but Dave was very disappointed when he came home.

"It was in a little old lady's backyard. She had just the one colony and a beekeeper came and managed it for her. All she knew was that there were three kinds of bees, a queen, workers, and drones. Shucks, I know more than that from a book I read."

But his interest had been increased by seeing the hive and the bees at work, and he wanted a colony. My husband had died during a flu epidemic when Dave was six and I was hoping to return to Denver, which I did two years later, so we had to postpone getting bees.

We arrived in Denver the day before Thanksgiving when Dave was in the fifth grade, and in the spring he began talking of getting a colony of bees. He searched the Sunday paper for an advertisement of bees for sale.

"I found one!" He came running with the paper and spread it before me. "See. Can we go and buy it?" he asked eagerly.

The address was only a few blocks from our home, so we got in the car and went to see about buying bees. The man, elderly and of foreign descent, was very friendly.

"So this is the young man who wants to be a beekeeper?" he said as he took Dave's arm and smiled at him. "When did you decide on that?"

"A couple of years ago when I saw a colony in someone's backyard."

"Now you can see some in my backyard," the man said and led the way.

Dave was delighted with both the man and the bees, and we watched them for a while.

"How much are they and when can we take them?" Dave asked.

"Well, I'll let you have them for eight dollars, and you can have a stand to put the hive on, also a super for some honey. But you will have to buy some frames to put in it and some foundation to put in the frames."

"And when can we take them?" Dave asked eagerly.

The man rubbed his chin and smiled at Dave, "You can take them today, but you will have to wait until the sun sets or there will be a lot of field bees that won't get home."

I looked at my watch; it was three o'clock. "We can come for them after seven, if that is all right."

"Fine," the man nodded. "Now for a few instructions. You will have to open the hive to see how the bees are doing. I haven't done anything for them yet, so I can't tell you."

Dave's face lit up with interest as he asked, "How do I do that?"

"You have a big knife?"

Dave nodded, a bit puzzled.

"You take the cover off and cut the frames apart so you can pick them out and see the brood; if there are a lot, the queen is a good one. There will be eggs in some cells and larvae and sealed brood in others. You'll need some bee gloves and a veil so you don't get stung, and later on some tools, a frame grip, and a hive tool."

Dave asked, "Do you have more bees than these?"

"I once had about a thousand colonies, but that was some years ago. Now all I have are these three and they are all for sale. Which one do you want?"

Dave walked around looking at the three hives, and at last he said, "I think I'd like the ones in the silver hive."

So that was the one we bought. I counted out the eight dollars and we said, "See you after seven."

When we got home Dave fixed a place in the backyard under an elm tree to set the hive and at seven we went back to get it.

"The bees are all carefully tucked in so none will get out as you drive home," the man said. "I stuffed a strip of old sheet in the opening along the bottom of the hive where their door is. When you get home and set them where you want them, just put a board slanting up in front so they will know they have been moved, pull the cloth away, and leave them for a day or two." Then he helped Dave put them in the trunk of the car.

Dave closed the trunk, said, "Goodbye, and thank you," and we were on our way home. We unloaded the bees and put them on the platform under the tree. Dave was careful to put up the board in front and pull out the piece of cloth. Then he brought the super in the house, and left the bees to become acquainted with their new surroundings.

The next day after school we went to Ward's and bought a hive tool, frame grip, bee gloves, and veil for six dollars and fifty cents. Dave insisted on keeping track of costs so he would

Tools

smoker

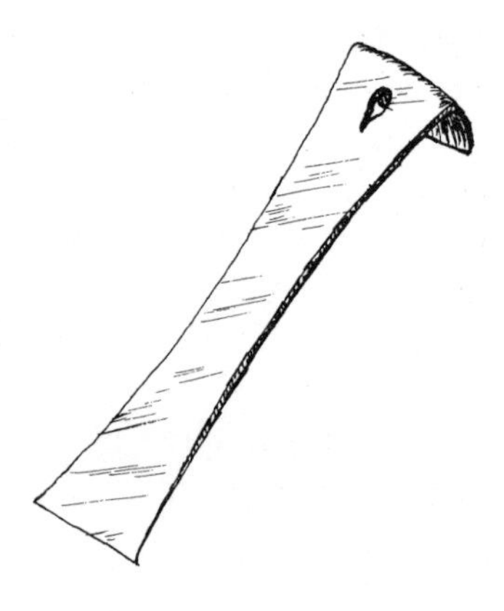

hive tool

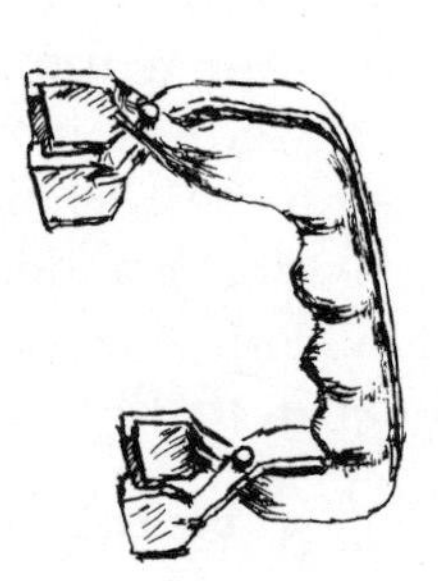

frame grip

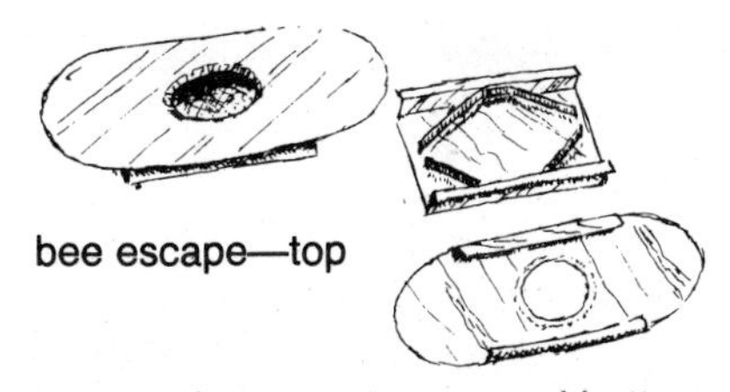
bee escape—top
inner spring passageway and bottom

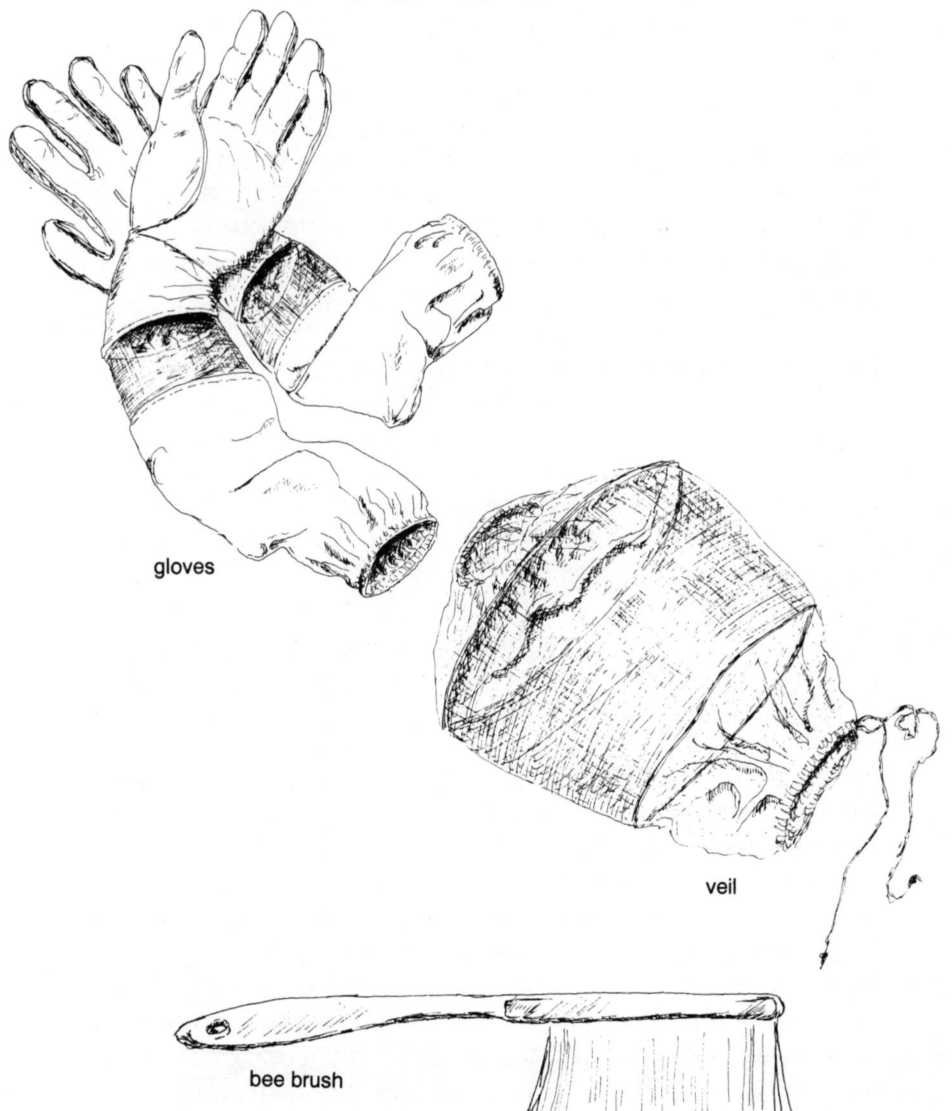
gloves
veil
bee brush

know if he ever made a profit. "I can hardly wait three days to open the hive and see what's inside," he said on the way home.

So on Wednesday when he got home from school, he put the veil around an old hat, tied the drawstring around his turned-up shirt collar, and with the gloves on he removed the hive cover. With my big carving knife, he began cutting the frames apart, but he must have also cut some bees, for the air was filled with angry ones. He was not stung but I was, even though I was standing about twenty feet away watching the proceedings.

Soon there were bees in my hair and I was stung about fifteen times on my face. I ran into the house, brushed the bees from my hair, and removed the stings that were in my skin, all fifteen of them. Then I bathed my face in cold water.

By this time Dave came hurrying in. "I closed the hive. Are you all right?" He was pale and rather frightened.

"For the time being, at least."

He was trembling with anxiety, "Gosh, I never thought they would be so mad. I did just what that man said to."

Later I went downstairs to Dave's hobby room to tell him I was going to put the car in the garage, and much to my surprise I saw two of him. In fact I saw two of everything in the room, but I felt normal though a bit tired. In the morning my face was so swollen that Dave was scared when he saw me.

"You look terrible! What can we do?" he asked.

"Nothing now, just wait for the swelling to go down." I had not looked in the mirror yet, but now I did. No wonder Dave was scared, my face looked like a football. I was scared too and called the doctor.

He wasn't much of a comfort: "There is nothing to do now but wait for the swelling to go down. But be careful not to let that happen again; you could have died last night."

But I hadn't died and that was what was important. I stayed in the house or yard for almost a week until I looked normal. I began to rationalize the situation,"Now that we are in the bee business, I think we better get a book on beekeeping—somehow I don't think that was the way to open a hive," I told Dave.

"I don't think so either, but that's what the man said. I wonder how he did it without getting stung by mad bees."

Thinking back to mother's beekeeping I couldn't remember

how she did it. Then I realized that she never opened a hive, since hers had no removable tops or frames. They were just a box with a queen excluder slot in the top. All she did was to remove the small box with the honey in it. It must have had worker bees in it because I remember that she let it stand open on top of the hive until after sunset so they would return to the hive.

However, we intended to be scientific beekeepers and that meant checking the hive frames for quantity of brood and queen cells and rating the queen as good, poor, or indifferent. So we bought a book on beekeeping and after Dave read it he said, "We will need a smoker to use when opening a hive. Smoke causes bees to gorge on honey and makes them docile like a swarm that has just left the hive. Some beekeepers say that smoke scares them, so I'm all for it."

A smoker is a cylindrical firebox with a bellows attached to one side, a spout on the other. Dave lighted some rags, put them in the firebox, closed the lid, and pumped the bellows to get a cloud of smoke. He pried up one end of the cover, pumped smoke under it, then lifted it off to expose the frames.

This time the bees were quiet; whenever they showed signs of becoming excited as he worked, he pumped in a little more smoke. Using the hive tool, he cut the propolis that held the frames together and carefully pried up each end of a frame until he could grasp it with the frame grip and slowly lift it out.

Propolis is a bee glue with which they cement everything together in the hive.

The frame was beautiful, full of eggs, larvae, and sealed brood.

"At least the queen is good," Dave said, "and look at all the bees working all over the comb. They are tending the brood and putting honey in the cells."

Dave inspected all the frames and the bees were quiet. This was quite a different performance from his first one. The book had made a knowing person out of him and he worked almost like a professional and was on the way to becoming a good beekeeper.

Bees

There are many kinds of bees. Henry Friese estimated there are as many as 20,000 kinds, different in structure, color, and size. There is a tropical bee so tiny that it was described from a specimen that lodged in the eye of an entomologist. At the other extreme are the giant bees of the East Indies, an inch and a half in length.

Most bees forage during the day, but there is a tropical bee that visits flowers at night. There are solitary bees and social bees; the latter class includes the honey bees. They are highly developed, having four membranous wings, of which the front ones are large, and metamorphose through larva to nymph.

The queen of social bees has a longer life span than her descendants. She shares the nest with her progeny, who love her, feed her, bring in the stores, tend the brood, and keep the home clean.

The skeleton of a bee is not internal but an external horny substance called chitin, which is covered with fine hairs. The body is made up of head, with antennae, mouth, tongue, and eyes; thorax, with wings and legs; and abdomen, containing the honey sack, bowels, breathing organs, and sting.

On each side of the head is a composite eye and on top, three convex eyes that are microscopic. They are in the shape of a triangle and function in the hive so the bees can work in the dark. The organs of smell and feeling are located in the antennae; if a bee loses its antennae it will die. Bees communicate with their antennae, or vibrate their wings or hum, which draws them together in the dark.

The thorax consists of three rings, each bearing a pair of legs on the underside, and the last two, a pair of wings on the upper side. On the bee's left and right side the front wing is larger and hooks to the smaller one in flight, as the bee's body is bulky and must have large wings to fly. When the bee is not in flight, the wings are released and overlap; then they are small enough to allow the bee to work in the cells. The leg is composed of several joints, the last one having two claws with which to cling to objects. Between the claws is a pocket which secretes a sticky substance so the bee can cling to smooth surfaces by turning the claws back and using these pads.

The sting, at the end of the abdomen, is a very complicated organ which is not visible unless in use. Its auxiliaries are a poison sac the size of a small mustard seed and a firm, sharp sheath which supports the sting. The sting consists of two chestnut-colored spears of a horny substance. Each spear has nine barbs like fishhooks which anchor into the flesh for one-twelfth of an inch to prevent withdrawal. The poison is similar to snake venom or formic acid but more potent, and causes swelling, red blotches, local pain, and fever. These will pass in a few moments if the body has an immunity, otherwise they may hold for days. Twelve or more stings can cause nausea and collapse.

The bee cannot withdraw her sting unless it went in perpendicularly and she can turn around the wound, thus rolling up the barbs so they can come out. Otherwise when she breaks away she loses not only the sting but some intestines and soon dies. Occasionally only the poison sac and sting are lost and she may live a while. The sting, a terror to many, is indispensable to her protection. Without it this precious insect would long ago have been destroyed for her honey.

The honeybee is not native to America. The Spaniards brought these bees to Florida in 1763; they appeared in Kentucky in 1780, New York in 1793, and crossed the Mississippi in 1797.

Bees

worker

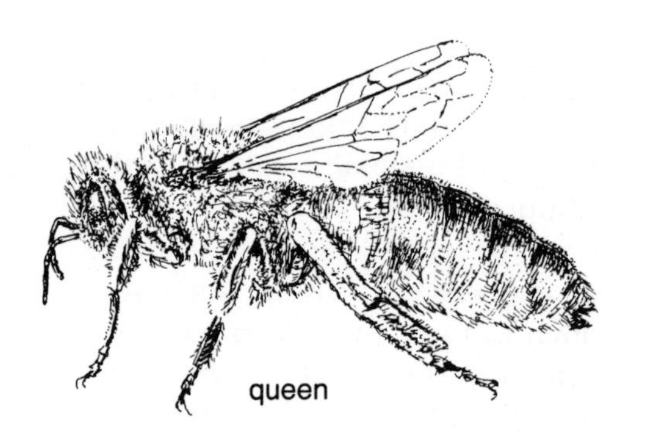

queen

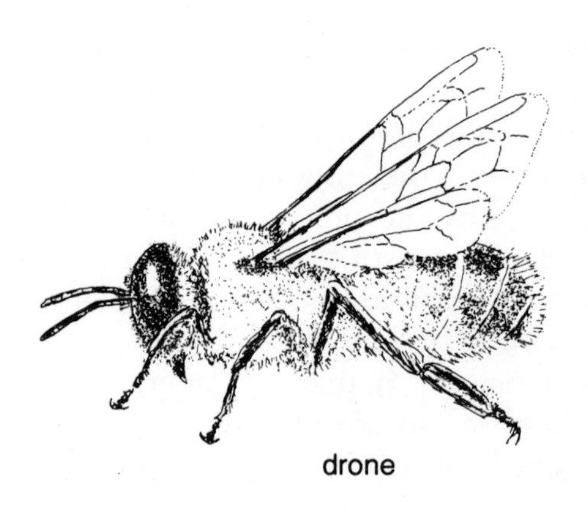

drone

Bees antedate man by many thousands of years, as geologists have found fossils in rock strata of the Stone Age. Early in history man recognized the bee as an unusual insect; he painted pictures of bees in caves in Spain as long as fifteen thousand years ago. One shows a honey gatherer climbing to a wild wax hive, while bees fly around. Bees were depicted on obelisks and temple walls in ancient Egypt and Babylonia, where beekeeping was first practiced; there is a relief in the Temple of the Sun (2000 B.C.) showing beekeeping in Egypt. Today bees are more widely distributed over the world than any other insect.

There are three types of bees: workers, queens, and drones. The worker is the smallest in size, about five-eighths of an inch long, but makes up the bulk of the hive population. In the fall and winter months there are usually twenty thousand, but by May there is a great increase to sixty thousand or more. Then swarming occurs, reducing the population in the hive to twenty or thirty thousand again. Before the honey flow the worker population increases rapidly and remains high all summer; workers live only six weeks at that time, literally working themselves to death. Workers hatched in September or later, when work is mostly over, will live for six months, or until well into spring.

Workers are female but do not lay eggs unless the queen is missing; then only drones will hatch because workers' eggs are not fertile. An egg destined to become a worker is laid in a cell of the comb by the queen. It is smaller than a grain of rice and hatches into a larva in three days. This larva is promptly surrounded by royal jelly, a whitish substance secreted by glands in the heads of the workers that tend the brood. After three days it is fed a combination of honey and pollen called bee bread, for five days. The larva is now large and fat and fills the cell, which is sealed by a thin cover of porous wax. It spins a silk cocoon from glands in its head, becomes a pupa, and undergoes the changes that make it a bee. On the twenty-first day after it was laid in the cell, it emerges a young bee.

The newly hatched nymph is small, pale gray, and very weak; but after several days she is normal in size, bright-colored, and covered with down which she loses with age and work. She remains in the hive for the first three weeks of her life; the first two are spent cleaning the cells, feeding larva, and attending the queen. On the eighth day she makes her first venture out of the hive, walks on the landing board, and observes the location of the hive by flying in circles around it (she soon tires and returns to the hive). Large numbers of young bees take part in these orientation flights around the hive at midday. These playful flights are a joy to watch and are sometimes mistaken for raiding parties.

When she is fourteen days old her royal jelly glands begin to fail but her wax-making glands have developed and she produces wax for combmaking. At twenty-one days she becomes a field bee and until her death brings in nectar and pollen or propolis in the baskets on her hind legs. She becomes shiny as the fuzz wears off, her wings become ragged from flying, and at length she is unable to return to the hive. Her life is only six weeks in the summer but more later on when the nectar flow is over. She is the bee with which we are familiar as we see her on flowers, fruit trees, and even on clover in lawns. Bees of all ages receive nectar, pollen, and propolis from field bees and store it in the hive.

A field bee may make ten nectar flights in a day, depending on the distance to be traveled. Pollen is easier to gather, so forty-seven flights are possible. A single bee brings in about one teaspoon of honey in her busy six-week life and she brings in pollen equal to her own weight.

To sum up, field bees bring in all the necessary products for the life of the colony in the hive. Young bees tend and feed the queen and brood, store the nectar and pollen, keep the hive clean and cool, and do the stinging to protect the colony. When a worker stings another bee she survives, but when she stings a person she dies in a matter of hours. Each bee can do the work of any other if need be; age is not just a matter of time but is related to bodily functions and their use. The critical time is the nurse age and the production of royal jelly in the glands of the head. Ducts open on the back of the tongue and the food flows forward in the feeding of larvae or queen.

When the three-day-old bee begins nursing larvae, the amount of work depends on the number of larvae and nurses. In late spring when the demand is great, every nurse will work until her glands deteriorate—then she turns to making wax. A worker hatched in October may not use her royal jelly glands until spring when brood rearing begins.

Full development of poison doesn't take place until the field-bee stage; guard duty is done stationed near the entrance or on the lower edges of the combs. When a guard is alerted its sting will protrude and a tiny drop of venom will appear at its tip. Its pungent odor will spread through the hive, the normal hum changes to an angry buzz, and many bees are alerted to attack. When a colony is thoroughly alerted it may remain so for twenty-four hours and woe to any one who comes too close to the hive.

The queen was first discovered by Butler in England in 1609, but the fact was ignored until 1737 when by dissection Swammerdam ascertained her sex. Prior to this the large bee in a colony had been called the king bee.

The queen bee is indispensable, since she lays the eggs that produce the successive generations of workers. She is the only perfect female; laying eggs is her only function in life. Her body is long and tapering, but her wings are shorter in proportion to her body than workers' wings. Her underbody is lighter in color and the upper part is darker than that of the workers. She moves more slowly but can hurry if alarmed. There are no sunny hours in the fields for her but no exacting work in the hive either.

A good queen lays as many as three thousand eggs a day during swarming time and the summer honey flow, but only a few in the fall. Three thousand eggs a day may seem an exaggeration, but is easily explained. In a hive there are about 2,400 square inches of comb occupied by brood, with 25 cells to the square inch, making 60,000 brood cells. The brood develops from egg to insect in twenty-one days, so the queen must lay close to three thousand eggs each day to keep the colony at maximum strength. This profuse laying exhausts the queen and she becomes an enormous eater, so her attendants must be constantly feeding her.

Oddly enough, the egg destined to become a queen is no different from that intended to become a worker—which is what it would become without the special feeding. This larva is fed great quantities of royal jelly and no bee bread; this is the one thing that makes the difference between a worker and a queen. And what a difference! The organs of reproduction are completely developed, her size and color are changed, the lower jaw is shorter and the head rounder, the abdomen lacks wax-making glands, there are no pollen bags on the rear legs, the sting is curved and one-third longer than that of the worker.

But these physical differences are not all; her instincts are also changed. Workers thrust out their stings on the slightest provocation, but a queen may be injured and still not sting. A worker has great respect for a queen, but a queen will destroy another of her kind on sight. A queen leaves the hive only for fertilization flight or to go with a swarm and she lives for five to six years while workers live six weeks to six months. When a queen becomes old she is usually not a good layer and the bees will supersede her. In commercial apiculture she is replaced every two years by the beekeeper to ensure a strong colony.

Bees create a new queen if the colony is crowded and the swarming impulse is on in the spring, if they lose her by accident, or if she is failing because of age. To create a new queen for swarming, eggs are laid in special cups, usually on the bottom of combs, and treated as potential queens from the start. As summer approaches, these cells are built and eggs laid in a number of them on successive days to ensure success. The queen doesn't voluntarily lay in these cups, but is forced to by her attendants; as the eggs hatch and the larvae grow, the cells are extended downward until they are about one inch long.

When bees supersede or lose their queen there must be eggs in the comb or larvae less than three days old because after that they are being fed bee bread, which a queen must never have had. The workers construct a queen cell around an egg or larvae in a worker cell on the face of the comb. The building proceeds outward and down until it looks the same as a swarming cell and the larva enters it. The workers may also transfer an egg or newly hatched larva from a common cell to a queen cell, which is much larger and resembles a small peanut.

The egg hatches into a larva in three days and is fed great quantities of royal jelly until it is sealed in its cell on the sixth day and remains there for seven days. On the sixteenth day the adult queen emerges, five days sooner than a worker. She remains with the colony if she is superseding an old queen (which the workers have killed), or if the older queen has left with a swarm.

The newly hatched queen is weak and fuzzy, but like the workers soon gains strength and starts over the combs in search of a rival. If she finds one, either hatched or still in the cell, she stings her to death and the workers throw the rival out of the hive. If the rival is still in the pupa she doesn't sting her but the workers open the cell and pull the rival out. Occasionally two queens may hatch at once, in which case they rush at each other and duel until one receives a fatal sting and dies. The victor is uninjured, for there is one fatal thrust between the abdominal plates and death is swift to the loser. The workers take no part in the battle. There is a slight chance that both queens will be killed; then the next queen to hatch will be guarded carefully.

The virgin queen begins to practice her nuptial flight when she is two or three days old. She flies cautiously at first, circling the hive and returning in minutes. Each flight strengthens her wings and succeeding flights circle wider. She must learn the location of the hive, for she would be instantly killed if she entered the wrong one after her mating flight. Queens usually mate between the fifth and tenth day; they must mate before three weeks or their mating capacity is lost. She can still lay eggs, but like the eggs of a worker they will not be fertile and only drones will hatch, so the colony will be doomed. The beekeeper must intervene to replace her, as there are no larva or eggs for the bees to use to replace her. She will begin laying forty-eight hours after mating.

Even a fertilized queen can lay infertile eggs at will and does when she comes to an oversized cell needed for a drone. These the bees plan so there will always be drones in the colony.

When a queen goes on her mating flight all the drones from her colony and any adjacent ones follow her. She flies very high and very fast and only the fastest drone overtakes her. This is survival of the fittest, only he doesn't survive. Like a worker stinging a person and then dying because some of her entrails remain with the sting, so the drone loses his genitals and some of his entrails when he breaks from the queen, and dies.

After mating the queen returns to the hive, followed by the unfulfilled drones. Occasionally she doesn't return, having been caught by a wasp and taken to its nest to feed to its young, or eaten by a bird. She may even enter the wrong hive if the hives are spaced too close or are too uniform in size and color. If this happens to the queen of a new colony they are in desperate straits and the beekeeper must supply material for the bees to create a queen. He may take a comb containing eggs, larvae, and sealed brood from an established colony and the bees will make a queen in about ten days. Or he may give a comb with queen cells on it and there will be a new queen in about a week.

Some writers on bees maintain that the queen never leaves the hive after her mating flight except to swarm, and that the thousands of eggs she will lay are all fertilized by one mating which is with as many as ten drones, five in the first half-hour flight and the rest in second and third flights that occur on successive days.

But a Canadian scientist investigated the number of flights a queen makes. He daubed one hundred queens with a nail polish containing radioactive zirconium and placed Geiger counters under the hives. On entering or leaving, the queens passed over the counters, recording an average of seven flights each. The question of why a shy queen would leave her colony for other than a mating flight was not answered, however.

Actually a queen stays in the hive except for mating or swarming flights, since her wings, being small for her body, do not lend themselves readily to constant flight. If a queen is to be taken from her colony she must be handled very gently and the bees will not sting. She must not be allowed to fly, as she might become lost. Having held a queen loosely in my closed palm, I am greatly moved by the gentleness of this insect that is capable of inflicting so much pain.

There is a striking difference between an old queen and a virgin. An old queen has a large rounded abdomen and moves with matronly dignity; her body has lost most of its fuzz from the caressing of her adoring children. A virgin has a small slender abdomen and moves with quick steps that lack dignity; her body has all the fuzz it had when she hatched. A colony that has had an impregnated queen will not readily accept a virgin except in an emergency. They may "ball" her (keeping her captive in a compact cluster), so it is best to give the colony a comb with a queen cell and let them make their own. Consciousness of a queen's presence binds the colony together and keeps the organization running smoothly. Her great capacity to produce eggs plus her inability to gather food, build comb, or feed offspring render her completely dependent on the workers who feed her mouth to mouth, although she can take food from cells if necessary.

A queen begins to lay two days after her impregnation and the eggs are laid in cells cleaned and prepared by the workers. She looks in the cell, moves forward, grasps the edges of nearby cells, arches her back and inserts her abdomen into the bottom of the cell to deposit an egg which attaches itself by a sticky secretion. It stands vertically, is white, about one-sixteenth of an inch in length and easily visible. An egg is laid every few seconds with intervals of feeding and grooming, for twenty-four hours a day through spring and summer.

Drones are male, larger than workers, with fat shiny bodies about three-fourths of an inch long. They have no sting, no proboscis to gather nectar, and no pollen baskets on their legs. They are raised in spring several weeks before new queens, as they take longer to mature. The queen passes an egg into a drone cell without benefit of sperm and a male is produced. He has no father, so he inherits the characteristics of the queen; this is called parthenogenesis. The larva is fed royal jelly for three days, then pollen and honey for seven days, and sealed in its cell. He emerges on the twenty-fourth day and needs fourteen more to reach maturity. He has a large hairy head, large antennae, and large compound eyes that almost meet on his face and extend around to the back of his head. The 8,500 facets in each one make almost 360-degree vision possible. But compound eyes are not good for detail and on mating flights he may chase anything that moves across his field of vision.

Drones appear in the hive in April and are to be found until after the honey flow at frost. All summer they swagger around in the hive, demanding service from their small sisters who pamper and groom them for the mating flight. They are accepted in any hive during their potency, so they may drift as far as seven miles. In this way nature maintains crossbreeding for the vigor of the species. They gather in groups out of the way of the workers and gorge on honey. In the fall justice descends and the atmosphere of the hive changes: the workers become belligerent and attack the drones, who have no means of defense. They are kept from food, stung to death and removed from the hive, or driven out to die of cold and hunger. The bees cannot afford to feed them during the winter, for one drone consumes enough honey to feed five workers and these indolent insects are no longer needed. Drones who don't mate live several months at most, only to meet this inconsiderate treatment and become food for birds, mice, or toads when tossed out of the hive.

A drone is about the most complete freeloader in nature, living in luxury without working; yet if the drones are removed from a hive early in the season the colony is restless, paying no attention to the queen in the hive, either before or after mating. The sole function of the drone is to fertilize the queen; death is his reward for success. Fertilization of the queen occurs high in the air because the sex organs of the drone cannot be extruded unless the abdomen is swelled by air in the trachea, which happens only during swift flight, and this movement cannot be reversed. The drone mounts the queen, copulation takes place, and the drone falls dying to the earth. Even if several drones mate with the queen, the vast majority never fulfill their life purpose.

Why are so many drones allowed to live in a colony all summer when only a few are needed to fertilize queens? It is for the protection of the colony, as the queen might die or become unproductive and have to be superseded. Then too, drones are large and easily seen by predatory birds or insects that might catch the queen on flight if she were not accompanied by a great number of drones. As it is, they can grab a fat drone instead.

There seems to be a great misunderstanding about drones and the creating of new queens. In an article on ecology the writer stated: "Bees have a way of perpetuating themselves. In their society there is a method of producing a queen when needed. An ordinary slouch of drone is chosen and by feeding it royal jelly, something you can't buy in the supermarket, they completely change its anatomy and give it the ability to lay eggs."

Nothing could be farther from the truth. Regardless of how

much royal jelly might be fed to a drone he would remain a "slouch of a drone," or he might die from being overfed. Even an egg intended to become a drone or a drone larva can never produce a queen. Only an egg intended to become a worker is used and the feeding of royal jelly produces a queen.

However, there is something mysterious about drones. Hatching from unfertilized eggs that the queen or workers may lay, they possess the power of life and can fertilize the queen for her life of egg laying to renew the colony. Bees have been known to try in desperation to secure a queen by feeding royal jelly to a drone larva, but it died from overfeeding, in its cell.

The Colony

Life in the colony is well organized. It was long a mystery how the work was conducted: some thought the queen directed everything, others thought each bee, like a human being, had a definite trade all its life. Dr. G. A. Rosch conducted a study in the 1920's that laid the foundation for our present understanding. Using a glass observation hive and marked bees, he determined the age and duties of workers. He found that the newly hatched worker may start its field work as a water carrier; water is necessary in brood rearing. A strong colony may use a pint daily and water is carried in the honey stomach. The hive bees take water and distribute it where needed. Water should be in or near the apiary to conserve the life and energy of the bees.

Nectar and pollen are gathered whenever available: some bees gather one or the other, others gather both. To gather nectar the worker has a special proboscis; in the center is a long slender tongue which can be extended or contracted. When not in use it is curled beneath the mouth but extends three-sixteenths of an inch beyond the jaws to collect nectar. From the mouth the nectar passes through the thorax in a long tube to the honey sac at the top of the abdomen. The field bee gives the nectar to a hive bee, which stores it in the comb.

Each bee must be able to recognize other bees of the colony in order to detect strangers who might steal food. This is done by smell. There is an odor common to bees of a colony, so pronounced that when a swarm leaves a colony and settles nearby, in a few days the bees are antagonistic. The colony odor is maintained by reciprocal exchange as bees share food in the hive. Bees sometimes enter the wrong hive by mistake or are blown off their course; they are stung to death unless they are laden with nectar or pollen.

The language of bees was first observed by Karl von Frisch; it has since been confirmed and accepted by scientific investigation everywhere. Bees have very high communication levels and this accounts for the efficiency of the colony. It is amazing that this takes place in a brain the size of a grain of millet. Frisch put a dish of honey about twenty yards from a hive; and while it took a few minutes for the first bee to find the honey, just moments after the bee had returned to the hive several more arrived and in half an hour there were more than one hundred at the dish.

Most field bees do not search for food; there are scouts for nectar and pollen. When food is found the scout returns and performs a dance on a comb, telling distance, direction, quantity, and source of nectar to be had, according to the gyrations. Distance is told by the speed of the dance, probably denoting the energy required to make the flight. The longer the distance the slower the dance. The dance circles may be small or large or she may rush across the comb in a straight line. A couple of seconds of this indicates a distance of half a mile. If food is two hundred yards away there may be seven complete dance circles in fifteen seconds, if four miles away, only two circles. The simple round dance describes food nearby and the bee runs in small circles in one spot on the comb, alternately to left and right. A more precise dance ending in a "wagtail" (figure eight with two circles made in opposite directions with a straight run between) shows distance. It is the straight run that gives the clue to the distance.

The greater the quantity of nectar the more excited the dancer is and the more vigorous the dance as she wiggles and waggles her abdomen, and the more recruits are obtained. Occasionally she stops and gives samples of nectar to the watchers, who touch her with their antennae to determine the flower. When she leaves, many accompany her.

There is also a pollen dance where the bee with the pollen turns a half circle, runs back to the starting point, turns another half circle, then goes to the starting point and wags her abdomen. This is done a number of times before she leaves the hive with many bees following her to bring back the pollen which is needed for the brood.

If the workers are too few to care for the brood or there is a lack of bee bread, the queen stands on the comb and extrudes her eggs, which the workers devour. Sometimes bees fail to cap brood cells, but the larva hatches anyway as it spins a cocoon at the mouth of the cell in about thirty-six hours. When the larva emerges from the cell the cocoon remains as a lining, so brood cells become smaller and stronger as they change tenants; however, the cell walls are so thin that they are little altered in twenty years of brood rearing. Occasionally the bees clean them a bit.

In foraging for food, bees fly a radius of several miles in all directions, at ten to fifteen miles an hour. A bee has two stomachs, a personal stomach for her food and a honey sac where she stores the nectar as she forages. Thirty-seven thousand flights of thirty to sixty minutes each are required to bring in one pound of honey. To do this the bees must visit half a million blossoms, all of the same type. It is more difficult on windy days, for at fifteen miles an hour their wings become tattered; this is why workers live only six weeks in the summer. They actually work themselves to death, for when they can no longer fly they die. But workers have been seen to crawl into an empty cell and rest for half an hour.

In their work of gathering nectar and pollen bees do 80 to 85 percent of all cross-pollination as they visit flowers. Other insects carry pollen, but they go from flower to flower indiscriminately, so that pollen that doesn't reach a plant of its kind is wasted. Bees do not intentionally play cupid for nature, but as they seek nectar and crawl into flowers the pollen drops on to their hairy bodies and is carried from flower to flower. They visit similar flowers as long as they are available, so the pollen is used effectively.

Bees generate heat in the hive in cold weather by clustering on the combs and using muscular activity such as shivering. They can be warm in the hive even though it is below freezing outside. This activity requires much honey from their winter stores; if the honey over which they are clustered is consumed before the cold abates they will die, for they do not move to a new area of comb while clustered.

The ability to generate heat gives greater adaptability to climate because bees can regulate temperature and humidity in the hive. It is also important in brood raising, for which a temperature of 95° Fahrenheit must be maintained regardless of outside temperature. Most of the heat is generated by the metabolism of the brood and conserved by the mass of bees that cover it. Away from the brood the temperature may be much cooler as adult bees work in as low as 57°. In hot weather hive bees cool the hive by fanning their wings and groups are placed to direct the air flow to best advantage. Many are on the bottom of the combs; others are on the alighting board facing outward to send fresh air inside, or facing inward to draw stale air out. The roar of their wings may be heard outside the hive. Humidity is controlled in the brood area, being kept fairly constant at 45 percent, for the digestive process creates water vapor. Any excess is removed by wing fanning. In hot, dry weather water is brought into the hive and spread on the dimpled comb surfaces or on parts of the hive to evaporate into the air current created by the bees. This cools the air and raises the humidity and is a very efficient system of air conditioning.

Bees have remarkable vision. By merely glancing at the sky they determine immediately the sun's position, the time of day, or the location of food. They fly on a sunbeam and never get lost.

While workers live six weeks to six months and queens live four to five years, this individual age of bees must not be confused with that of colonies. Colonies have been known to occupy the same home for generations, as long as twenty years. Some bees have been reported living in the same place for forty years.

There are many races of bees; those used commercially are the Italian bees because they are gentle and easy to handle and the most industrious. They still exist in a pure strain after two thousand years and are easily distinguished by their golden color. They are more resistant to disease, do not swarm readily, are less sensative to cold, and their queens are more prolific. If their hive is removed while they are in the field they will return and cluster at the exact spot, and some of the darker bees will hunt around until they find the hive.

Other races are Cyprians from Cyprus, which are vicious and hard to handle; Syrian bees from Palestine; and Egyptian bees, which are not much in demand in the Western world. Carniolan bees are from Austria and Caucasians from the Caucasus Mountains. They are quite similar—dark and gentle—but swarm excessively, which makes beekeeping quite arduous.

Hives

The first hives provided by people for bees were as crude as their natural abode in trees. It wasn't so long ago that "bee gums" were still in use. They were made from the trunk or a large branch of the gum tree, hollowed out with two sticks crossing in the middle. There was a rough board on each end with a notch at the lower end of one for an entrance.

In Europe hives were made of straw rope or willow, coiled in a conical shape two-and-one-half feet high and set on a board two-and-one-half feet square with a groove in the center for an entrance. A friend of mine who kept bees in Sweden used such a hive, and wrapped it in rags or an old blanket to protect the bees during the cold winters. These straw hives were called skeps and had a plug of wood at the top for removal of honey, which bees store at the highest point in the hive. In Sweden the skeps are not conical but flat on top with a three-inch hole over which the super, a wooden box, is placed and removed when full of honey. The beekeeper has no way of knowing the strength of the colony or the effectiveness of the queen, as he never looks in the skep itself. Today's beekeepers would feel quite handicapped if they could not check the condition of hives.

Hives

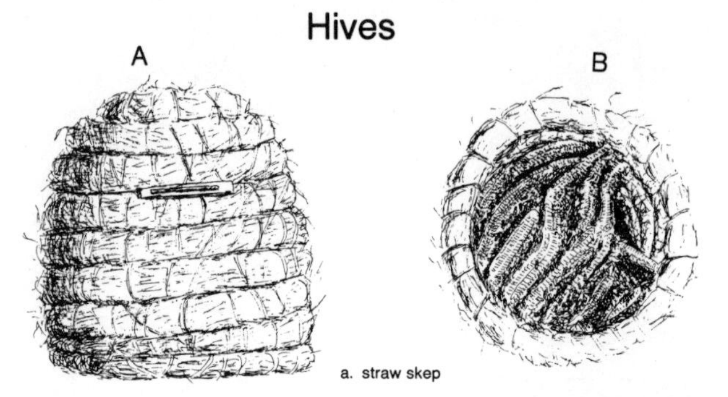

a. straw skep

b. top view with lid removed, showing comb without frames

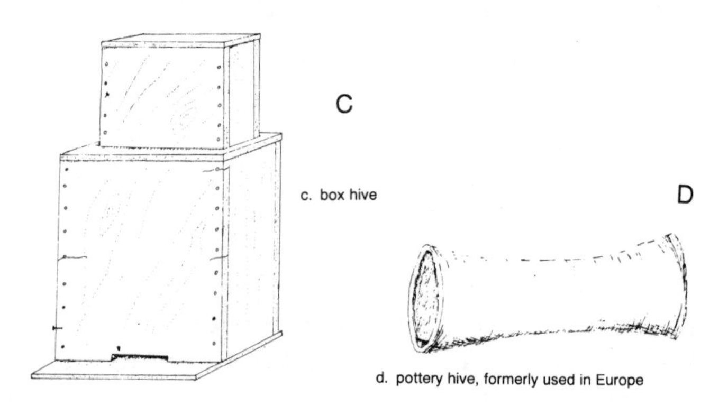

c. box hive

d. pottery hive, formerly used in Europe

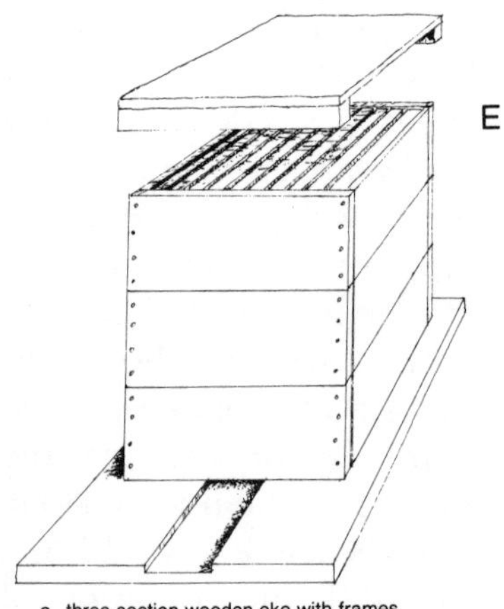

e. three-section wooden eke with frames

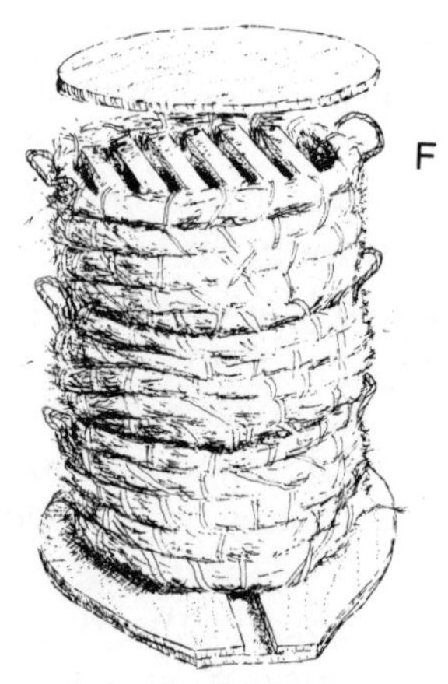

f. straw eke with frames

In addition there were pottery hives, which are still used in some areas of Asia and Africa. Pottery tubes are laid on their sides with each end closed by a wooden disk. The front one has a notch at the bottom for an entrance. The rear disk is removed to take the honey, which is stored at the back of the tube. These pottery hives were the most sensible of the old kinds.

In Greece hives were set in specially built thick stone walls with the entrance on the outside of the wall. Sometimes they were built into the walls of houses and the honey was taken from inside the house or wall, away from the flight of the bees.

In order to get honey from gums or boxes, the beekeeper sometimes drove the bees to another hive and took all the honey, but most of these impoverished colonies died. This led to killing the bees over a brimstone pit, as it was much easier than driving them out. This practice destroyed so many bees in the eighteenth century that Emperor Joseph II of Austria gave a florin per colony to all beekeepers who would cut out the comb instead of brimstoning the bees.

In brimstoning, the hive was placed over a pit where sulfur was burned and the bees tumbled from their home by the thousands, engulfed in blue sulfurous flames, and the beekeeper had all the honey. Incidentally, sulfur was what Jed put into his bucket of rags and paper when he killed the bees under mother's porch, but we did not know it then.

In 1830 a boy told of helping an old priest harvest honey from twenty colonies: "Clothed in heavy linen frocks, with wire masks strong enough to turn a sword, we cut the comb from the back of the hives with a curved knife. It was April, and we sweated under the scorching sun in our heavy garments. The crop harvested was light but the damage inflicted on the bees was immense, for they had to rebuild their combs at a time when queens begin their greatest laying."

Actually, beekeepers thought this was beneficial to the bees as it compelled them to make new combs, little realizing that this left no place for the queen to lay until the combmaking was accomplished.

Some beekeepers noticed that bees store honey in the highest part of the hive and added upper stories that communicated with the hive through a hole in the bottom. But when the story was deep or the hole small the bees would not store honey in it, so they replaced the top of the hive with slats or boards. When a story was full it was removed and replaced with another. These stories were later divided into several horizontal sections called ekes, and to cut the comb they pulled a wire through between the stories. Instead of using caps and getting only a small amount of honey, these stories brought a large yield. As many as four would be piled up.

Early in the nineteenth century, apiarists realized that something had to be devised to prevent the tangled mass of combs that filled most of their hives. They desired more honey that would be free from brood.

Accordingly, in 1821 Radoun put triangular bars in an eke hive, to which the bees could fasten their combs. These could be removed without killing bees. Mr. Hammet made a square wooden eke with movable frames, and other apiarists made divided hives that could be swung open like a book to remove the combs. These did not permit much study or manipulation and it became necessary to devise a hive whose every comb and part could be promptly and easily removed. This led to the development, in 1851, of the Langstroth hive which is used today the world over, with variations.

It has ten movable frames, in which the combs are built, so suspended from rabbets at the top of the hive that they touch neither the top, bottom, nor sides of the hive. Between the frames and at the hive wall there are three-eighth-inch spaces called beeways. By this device combs can be removed as needed, without any cutting, and transferred to another hive. This invention placed American apiculture ahead of that of all other countries.

Before Langstroth, an interesting type of movable frame hive appeared in Germany. It was made of straw and the frames were removed from the bottom, so that to examine the hive it had to be inverted. There was no separate compartment for honey.

Still earlier than this, Huber invented the leaf hive, consisting of twelve frames an inch and a quarter in width, hinged together so they could be opened and shut at will. This was called the closed and standing-frame hive; the greatest objection to it was the difficulty of fitting the frames back together without killing bees.

Shortly after a patent was issued to Mr. Langstroth, Baron von Berlepsch invented small rectangular frames which the bees filled with combs. This facilitated the removal of combs but the design of the hive all but cancelled out the advantage. It opened from the rear like a cupboard and if it was necessary to reach the last frame, every one of the others had to be taken out. It held three stories for brood and one for honey.

There were a half-dozen attempts to find an all-around satisfactory arrangement for the handling of frames in a hive, but Langstroth's arrangement soon proved far and away the best. It enables the apiarist to operate without killing a single bee, as both hive and super are bottomless and topless boxes that can be piled endlessly, as all they need is a cover and a bottom board.

Our hives give us perfect control of the combs and we can manipulate them as we choose. This enables us to exchange brood or honey and pollen as the colonies may need them.

Even with our modern methods and scientific equipment the bees often fool us and do things which books say they never do. We console ourselves with the knowledge that bees do not read books, but their wonderful wisdom is certainly God-given and inspires us with awe.

A satisfactory hive, aside from giving the apiarist perfect control of the combs and adequate manipulation, should afford the bees protection from extremes of heat, cold, and dampness. The accommodation of the bees should also be given some thought, to aid the work of the busy honey gatherers, who should not be compelled to travel through densely crowded combs with their burdens. The hive should permit the use of foundation in frames.

The frames in a hive should be fully nine inches deep, for the queen lays her eggs in a circle in order to lose no time hunting for cells. Shallow frames will break the circle and compel her to lose time. If the comb is only five inches deep, the largest circle possible contains less than twenty square inches of comb, with five hundred worker cells on each side. At every half turn she will lose not only time but eggs, for in the honey season eggs must drop, if not deposited in a cell. Loss of eggs means loss of brood and this means loss of honey at harvest.

For this reason the brood chamber should not be shallow,

since there would have to be two to accommodate the brood. Then the bees that cover the brood to keep it warm would also keep warm the lower bar of the upper frames and the upper bar of the lower frames and the distance between the two, with no benefit from it.

Langstroth was correct in using frames of greater length than depth, since these allow more room for storing supplies. They are 19½ by 9 inches and furnish 1,710 square inches of comb, containing 93,000 cells of which a good queen, laying 3,500 eggs a day, will keep over 70,000 filled with brood. The workers will fill the remaining 20,000 with honey needed for the brood.

The Hoffman frame is similar to that of Langstroth, but the frames touch in the upper third of their sides and are spaced to allow ten to the hive. These are the ones we use and the ones handled by most supply houses.

Many beekeepers succeed and harvest good crops with smaller hives. One way to offset the problem of a small brood chamber is to use two early in the season, but remove the second one when the supers are set and create a new colony with all of the frames, some of the bees, and a queen. Or the brood frames may be given to a weak colony. A large brood chamber may easily be reduced in size if need be, but a small one cannot be enlarged except by the addition of an upper story.

Another disadvantage of using small hives is the excess of natural swarming which they cause, since the main cause of swarming is lack of room for the queen. Then too, if hives are so small that the most prolific queens can't be discovered, it is not possible to improve the stock.

The distance from center to center of the frames can be varied from one-and-three-eighth to one-and-a-half inches in the brood chamber by measured spacing. It is better to have the greater distance because it facilitates examining frames or exchanging combs to help weaker colonies with brood or honey. It also gives more room between the combs for the bees to cluster in winter as well as a greater thickness of honey to sustain them.

What's more, the frames must be equally spaced in the hive to keep the combs straight, for movable frames with crooked combs are worse than no frames.

Besides, the width of the top bar has something to do with the amount of bridges and brace comb built by the bees between the brood chamber and the super. A wide top bar, leaving only a narrow space for bee passage above, will prevent to a large extent the building of bridges. Hives should slant forward when occupied, to aid in the carrying out of dead bees, draining out moisture, and preventing rain from beating into the hive. The frames, thus, should run from back to front instead of side to side as in some hives.

All modern hives meet these requirements and are set on movable bottom boards to facilitate cleaning when necessary. The bottom board projects forward about three inches to support an entrance block in winter and furnish an alighting board for returning bees.

If the grain of wood in the bottom board runs from front to rear, it will shed water better, although one Swiss apiarist prefers to make it with the grain running from side to side, contending that it fits better into the hive and there is less swelling. The bottom board is the first part of the hive to decay and a hive body will outlast two bottom boards.

We use the Langstroth hives but make the bottom boards with slats on three sides, thus affording an entrance in front. There is a spacing wire that may be used to space the frames at the bottom. It is bent to fit the bottom of each frame; if the hive is to be moved, it prevents the frames from jarring. Bees usually secure the frames at the top, but there is no way to secure them at the bottom except with this wire. The entrance should be five-sixteenths of an inch, which gives easy entrance to the bees but keeps out mice.

Another useful implement of the beehive is a straw mat, flexible and porous, warm in winter and cool in summer, placed immediately over the frames. A good mat will last as long as the hive and will absorb all moisture that escapes from the bees in winter.

Hives should be painted to make them last and give a neat appearance to the apiary. However, no dark colors should be used (they absorb heat), nor should all hives be painted the same tint, as bees may drift into other hives if they are too alike. A white base with tints of blue, pink, or green can be used unless there is shrubbery to give the bees a landmark.

Every hive should have a number on the back, in black and large enough to be seen at a distance. All operations at each hive should be recorded in a book under the appropriate number.

At one time there was a wide variety of opinion on the best material for hives, from the hollow trunk of the cork tree to straw. Straw is warm in winter and cool in summer, but the difficulty in making it take and retain the proper shape for improved beekeeping caused it to be abandoned.

Nearly all modern hives are made of wood; the lighter and more spongy the wood, the more like a straw hive it will be. Cedar, poplar, and especially soft pine are excellent materials.

Hives may be made from small pieces of lumber from a lumber yard, or second-grade lumber may be used to keep costs down. The parts can be made easily and cheaply by anyone who can handle tools. Since lumber is constantly increasing in price and scarcity, the only way to have inexpensive hives is to make them. The overall size of our hives is 16x20x9½ inches, width, length, and depth respectively.

The heart side of the wood should be on the outside of the hive. Looking at the end of each board one can see by the circles of growth which is the heart side. A board always warps away from the heart side; it becomes convex and the hive will not pull apart. If you have seen a beehive or even a box gaping open and separating at the corners while the middle is in line, the mechanic who made it put the boards on the wrong side out. If the heart side had been outward, the corners would have tended to curl inward and the hives would have close, tight joints even if exposed to sun and rain.

Our First Colony

As we discussed our plans for beekeeping, Dave said, "Let's go in for comb honey. It's the best and we won't have all the stuck-up mess of extracting."

"I agree, since you are to do the work, I wouldn't want you to be all stuck-up either."

Our colony was still in the backyard, and as it was an established one, we allowed the bees to store the spring nectar in the hive to feed the brood and build stores for winter. About the middle of June we went to a bee supply house and Dave bought a package of thin foundation and ten frames, which he would put together for the super. Foundation is a thin sheet of beeswax embossed with a hexagonal design upon which the bees build the honey cells.

Dave assembled the frames, put foundation in them, set the super, and we waited until September for the results. Then Dave donned a flannel shirt, bee gloves, a hat with the bee veil tied around the base of the crown and at the neck, and opened the super. This time he was careful to pry up the lid and use the smoker. The super was full of honey, about forty pounds of it, but we were disappointed because it looked and tasted like light molasses. We gave some to friends and relatives and used the rest.

Dave went for the book and in about half an hour he said, "I know why the honey was dark. There is such an assortment of flowers in the city and most of them are the variety that has dark nectar. Bees should be in the country where there is a lot of one kind of flowers, like alfalfa or clover. Then we will have mild, light-colored honey."

"Where can we find a place in the country to keep bees?" I asked.

"Maybe a beekeeper could tell us of a place," Dave suggested. "We would not need it until spring, since the honey season is over."

While we were thinking about moving the bees, a neighbor lost her cool when she realized we had bees in the backyard. One morning she said to me, "I don't like those bees around here. If my daughter should get stung it would kill her."

To say that I was surprised is putting it mildly, but perhaps it was true. People have been known to die from multiple bee stings or if they are highly allergic.

"Isn't she taking a chance just going outdoors or to school?" I asked. "There are bees in trees even in town and they fly several miles to gather nectar. Of course they don't sting when gathering nectar, but they could be stepped on in the grass by a barefoot child, or picked in a flower. Then they would sting."

"But your bees are especially dangerous because they are by the fence and there are so many of them."

"I'll see if we can find a place for them in the country," I told her. "We've been talking of it anyway."

But she gave us no time. The next morning a policeman knocked at my back door and asked, "Do you have bees here in your backyard?"

"Is it against the law to keep bees in the city?" I asked.

He grinned, pushed his cap back an inch or two, scratched his head, and said, "No, there is no law because bees can't be controlled. But if you continue to keep them here you won't have many friends. People are afraid of bees kept near their yards or houses. You did not answer my question, do you have bees here in your yard?"

"Yes, back of that far tree, but we are going to find a place for them in the country and it may take some time."

"Whatever you do is OK with me. There was a complaint and I was sent to talk to you and I have." He tipped his hat and left.

Winter came and went and all was quiet; then in the spring our colony let out a swarm that settled just over the fence in this neighbor's yard and on her rosebush. Dave called the fire department to find what to do about it and was given a beekeeper's number to call. The beekeeper came about three o'clock with an empty hive and took the bees from the bush, showing Dave how to do it. His name was Johns and he said he had fifteen colonies out east of Denver.

"Would it be possible for us to take our colony out there?" Dave asked. "This neighbor doesn't like us to have bees in the backyard and we aren't too fond of city honey. It tastes like molasses."

Mr. Johns laughed. "I know what you mean." He looked thoughtfully at Dave, seeming to be interested in so young a boy wanting to keep bees.

"I think perhaps you could put your colony where mine are. I'll ask my friend and let you know this evening."

He left the hive under the bush to get all the scouts and came back that evening and loaded them into the back of his truck.

"The beekeeper always gets the swarm in payment for taking it," he said as he smiled at Dave. "I spoke to my friend and he said it is just dandy for you to bring your bees out," and he gave Dave the directions to go there.

Here was the answer to our problem, but for the moment Dave was thoughtful. Then he said, "So the beekeeper gets the bees. Let's get some hives and take bees for the fire department. Then we will have more colonies this year, otherwise we won't because we lost the swarm that came from ours."

"That is quite an idea, but hives and supers may be expensive; maybe we can't afford all we would need. Beekeeping was to be just a hobby, wasn't it?" I asked.

Dave looked in our farm catalog and whistled. "They sure are expensive. Even if we buy them knocked down and I assemble them it will be about twenty-three dollars for a complete hive with super. Maybe we could make them, but do we have time? The swarming season is here; ours left yesterday."

He was so right; we had let spring almost slip by, busy with other things. Here it was the first week in May, the month to take swarms that would produce honey by fall. That very old saying, "A swarm of bees in May is worth a load of hay, a swarm of bees in June is worth a silver spoon, but a swarm of bees in

July isn't worth a fly," is really true. A swarm taken in May will produce as much honey as an established colony, a swarm taken in June will produce a small harvest, but a swarm taken in July will most probably have to be fed through the winter.

While we puzzled over our problem we bought a hive with frames and wired foundation, a cover and bottom board. Dave had to assemble the hive and frames and put the foundation in them, but he was a willing worker. He kept track of expenses to know if there would be a profit in the fall, so he entered fifteen dollars in his small book. This did not include a super since we would not need one for six weeks, but then it would cost another twelve dollars, including a queen excluder.

To say that Dave was astonished was putting it mildly. "Wow! We can never afford to have a dozen colonies at that price, can we?" His voice was tinged with disappointment.

"It is quite expensive for one colony, but since you called the fire department and are on their bee pickup list there is no telling how many calls we will get."

"We may be getting into the bee business in a hurry." Dave laughed in spite of his concern.

"Let's get some lumber and try making our hives like you suggested," I said.

"Groovy! But I'm afraid you'll have to do it. I can't saw straight," Dave confessed.

"I can saw a one-inch-thick board straight, and since that is what it will be, I think I can do it."

"There will still be the cost of frames, foundation, and queen excluders," Dave said. "The excluders are necessary to keep the queen out of the supers because we want comb honey."

"The frames and foundation will be less than four dollars and we would have that for each hive anyway," I said as I studied the catalog. "We won't need the queen excluders until we set the supers later on."

So we did some figuring and found that from a six-foot length of one-by-ten we could make a hive and from the same length of one-by-eight we could make a super. In terms of money that would be quite a saving since the hive and super would cost only four dollars, but in terms of my time and ability it would remain to be seen.

So we bought the lumber and I turned carpenter and made

five hives with covers and bottom boards. I measured the boards for correct length and marked them with a pencil and ruler to keep the ends straight, for if I wasted lumber I would be defeating my purpose of making inexpensive hives.

One thing I had not anticipated was a groove on the top of each end of the hive, to hold the frames. By dint of ingenuity and hard work I made them in the ends before I put the hives together. Dave painted the outsides white and we were proud of our work when the five hives were stacked ready for bees and we were ready for calls through the fire department.

With all this preparation to take swarms we had not yet moved our colony from the backyard, but the following night we decided to take it out where Mr. Johns kept his.

"This will be a blast if we make it," Dave said as he stuffed a strip of cloth in the entrance.

We loaded the hive into the trunk of the car and the bees left as they had come over a year ago. We set them at one end of Mr. Johns's hives and propped a board up in front so that in the morning the bees would know they were in a different place. Dave pulled the cloth from the entrance and hurried away, because even at night guards are alert.

It was a lovely place in the moonlight. There were big trees and Johns had a large watering tank nearby. We returned home, happy in the thought that we were about to become beekeepers.

Taking Bees

That summer we took five swarms, just enough to fill the hives. The first call for a bee pickup came in May when Dave was in school, so I took the information.

It was a woman who said, "I was told to call this number to get someone to take a big bunch of bees off my porch. Can you do it?"

"I'm sure we can. Where do you live?" I asked and she gave me her address. "When did they come?"

"This morning. How soon can you get here?"

I told her that we would be there about four o'clock. When Dave walked in I said, "Guess what? We have our first bee pickup."

"Groovy! We're in business!" He put a hive with top and bottom boards in the trunk of the car and took five frames with wired foundation in them, and the bee brush.

"I'm going to wear my gloves and veil," he said. "No laggard of a bee is going to sting me."

Bees don't usually sting when swarming, for the simple reason that before the swarm leaves the hive, the bees gorge on honey until their abdomens are so distended that they can't bend to sting. This great supply of honey is to last them until

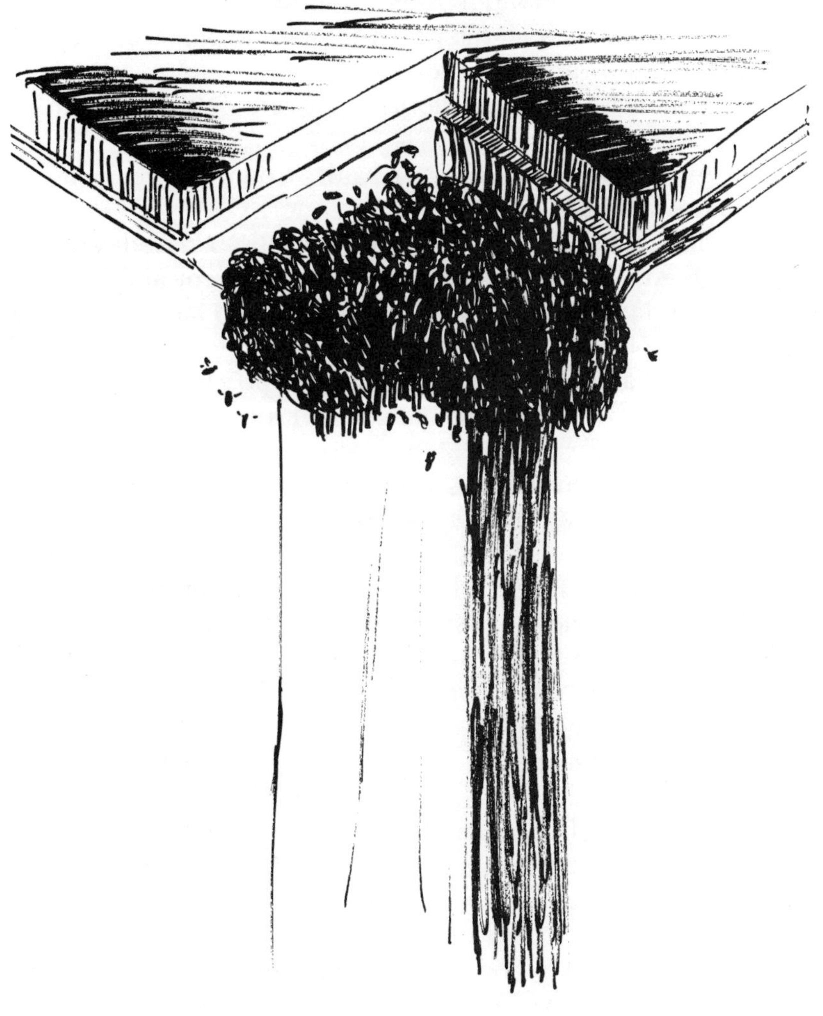

Swarm on porch pillar

they find a new home and start to make comb. However, there are usually several bees that made up their minds to go along at the last minute and didn't get to eat honey. They are hungry and grumpy and ready to sting anyone who bothers the swarm. Such a bee was what Dave had called a "laggard of a bee."

When we arrived at the house we saw a very large swarm clustered at the top of a porch pillar, so we went to work. Dave put the bottom board on the porch beside the pillar and set the hive on it while I rang the doorbell.

The woman came to the door and I said, "We've come to take the bees."

"Oh, for goodness' sakes!" she said. "You don't really mean it? I expected a fireman."

The idea amused me. "Uniform and all?" I asked, and we both laughed. "We represent the fire department and I assure you we will do it very well. Do you have a step stool or should Dave get up on the porch railing?"

She looked anxiously at the hive on the porch, but when she saw Dave in gloves and veil she seemed a bit more reassured and said, "I'll get a stool."

From then on Dave always wore his veil and gloves to create confidence. I resolved to get a set for myself so we would look real professional.

She brought the stool and asked, "Is it all right if I watch?"

"Of course," I said, "but stay inside the screen away from the bees. You said they came this morning?"

"Yes. I first saw them about ten o'clock and called the fire department right away."

"That's fine," I said. "The sooner we get to them the easier it is for us."

Dave got up on the stool and I handed him the bee brush and a frame. He brushed bees onto it, handed it to me to put into the hive, and I handed him another frame. Sort of an assembly line plan.

We worked quickly and silently until all the bees were on frames and in the hive. We knew we had the queen, since the bees were not flying around as they would be if the queen had run into a hole or crack. Of course there were lots of bees in the air during the transfer from pillar to hive but they were not trying to sting and they were going into the hive.

Dave put the cover on at an angle, leaving open spaces at the corners so the scouts and flying bees would enter the hive. As I gave the stool to the woman, she said, "My, that was easy. I thought it would take most of the day. Now what will you do with the bees? I hope you won't kill them."

"Oh no, we will keep them and put them with our other bees."

"Oh! You have more bees?" She gave a little shudder. "Well, I will say you are very brave."

"We would like to leave the hive on the porch until after sunset," I told her, and as she began to look worried, I hastened to explain.

"When a swarm lights anywhere, scouts go out to try to find a place for the swarm to live. Sometimes there aren't very many and again there may be enough to make a ball the size of a small orange. They had from ten to four o'clock to go out, so if we take the hive away now they will come back to your porch and cluster there. But if we leave the hive until after sunset, they will go into it as they come back and we will have all of them."

"Oh, do leave it," she said with relief. "I don't know what I'd do with a ball of bees up there. I'll tell the children to stay away from it."

When we went back that evening all was quiet, so I stuffed the entrance with cloth and Dave tied a rope around it very tightly. This was to prevent the hive from slipping off the bottom board as we drove home. There was no propolis to cement it on, since the bees had just occupied, and the one thing we didn't want was a trunk full of angry bees.

Before I could let go of my end of the hive I received a sharp pricking sting on my arm. There she was, the unprepared bee. She must have been sitting on the side of the hive, as none could have gotten out after I stuffed the entrance. I gasped and Dave asked, "What's the matter?"

"I just got stung."

"That settles it. You get gloves and a veil just as I have. I don't want you getting stung every time we take bees."

"I thought of that this morning when I saw how impressed the woman was when she saw you dressed for the occasion."

I bought bee gloves but the veil was not wire like Dave's. It was a mosquito netting hat from an army surplus store and had a taffeta crown. It was dark green (as it should be to see through) and very good. I still wear it.

The next morning Dave said, "Golly, I don't want to go to school today."

"For goodness' sakes, why? Are you sick? You look all right to me." I felt his forehead to see if he had temperature.

"I want to watch the hive to see the queen go for her mating flight. That's more interesting than school."

"You would have to stay right there all day, and she might not go out until tomorrow or the next day. And she might be an old queen and would not go out at all," I said. So he went to school.

We picked up four more swarms that spring and that was it. There were four other beekeepers on the list, so the fire department gave out our phone numbers in rotation. We now had six colonies out by Mr. Johns's, and I had to make five supers for our fall honey. I made them for six dollars, about a third of what they would have cost at the bee supply house. Dave was delighted, but it was not easy work for me, especially that groove to hold the frames. However, I was experienced this time.

We still needed frames, foundation, and a queen excluder for each super, the latter to keep our honey free from eggs, larvae, and baby bees. A queen excluder is a wooden frame, the size of the super, with what looks like a cake rack built into it. The small bars are just far enough apart to allow a worker to crawl through, but it excludes the queen; hence its name.

The queen cannot enter the super but the workers fill it with honey when the hive is full. This we harvest in mid-September after placing the supers in mid-June.

We check the supers every few weeks and as soon as one is filled we remove it and put on another. June to September is the big honey flow in our area; after that, what nectar is brought in, the bees keep to add to their winter stores.

By fall our six colonies out in the country yielded light-colored honey from alfalfa and clover. All honey in a frame is the same kind, as bees do not mix honey in a frame. But another frame might be darker. Each frame holds four pounds of honey, so a super usually yields forty pounds.

Bees in the Backyard

One day in early June Dave was home sick and he heard the roar of a big swarm as it lit in a tree just outside his bedroom window.

"Wow," he almost shouted, "come and see the swarm of bees that has just come into our tree. It must be twenty feet up there. If only I wasn't sick with the flu I could take them, but I guess they are too high for you to try."

I went to his room and looked out the window. Indeed, they were high up. "I wonder how I could get them." I sat on the bed and we brainstormed the situation.

"Even if I got on the ladder it wouldn't help much. It is six feet and my five feet four plus my overhead reach of two feet makes only about thirteen feet. Seven feet short."

"How about using my nine-foot bamboo fishing pole?" Dave suggested. "That would reach them, but there's still the problem of how to take them. You couldn't just scrape them off, they'd get mad and you might kill the queen."

"I know!" I said after a bit. "I'll make a circle out of a coathanger, tack a pillowcase around it, and fasten it to the end of the pole. I'll have a giant butterfly net and maybe I can get them in it."

Dave grinned broadly. "Far out!"

When I got my invention completed it looked like a giant butterfly catcher, but with muslin instead of net.

I waved it in front of Dave. "Bring on your eighteen-inch butterflies."

"Groovy!" he said, and laughed. "That should do it."

"Now that I have something to take them with, there is another problem. What do I do with them? All our hives are full." In my eagerness to make a net I had not thought about a hive.

Dave snickered. "Surprise! We still have one."

Where? I haven't seen any around lately."

"It's in the basement behind the furnace. Remember, we bought one, then you made five before we got a call, and we only took five for the fire department." He looked very impish for he knew I had forgotten about the hive we bought.

I brought the hive up, set it on the bottom board under the tree branch, and put four frames in it. This was always done whenever and wherever we took bees. They like the smell of foundation and are more willing to stay if the queen is with them. In fact we never had a swarm abandon a hive, although I know some beekeepers whose bees have.

I donned veil, gloves, and jeans which I stuffed into boots so I would not be stung if bees fell around me.

"I'm all set," I called through the window to Dave. "Are you watching?"

In answer the screen door opened and he stepped out on the porch, clad in pajamas and robe, looking very much like a hospital patient out for air.

"You better not stay there," I said. "I don't know how this is going to work. I may spill bees all over the lawn."

"That's why I came out. Maybe I can stop the thing from crashing to the ground if you can't balance it."

How like Dave. My heart rejoiced in him. I climbed the ladder to the top step, raised the pole with the pillowcase on the end, and gently brushed the cluster of bees with it. Almost the entire clump tumbled into the pillowcase. I hadn't expected to get so many bees on the first try and the whole thing went crashing down just as Dave had said. My wrists and arms were not strong enough to balance a good five pounds of bees suddenly dropped into the case. But Dave was quick on the pickup and had the pole at arm's length and waist high as he said, "Got 'er!"

I came down the ladder, took the pole, inverted the case over the hive, and shook it slightly. The bees dropped into the hive onto the frames. Without waiting I climbed the ladder again and brushed the case along the branch a second time. This time there weren't many bees, so I could balance the pole. I came down and shook them into the hive. There were a lot of bees in the air now and I said to Dave, "You better go inside. I'm going to give it one more try, because it looks like the queen isn't in the hive. Too many bees are leaving."

Once more I climbed the ladder and once more I brushed the case along the branch. I wasn't too hopeful, as there was only a very small cluster still on the branch, but I got it. Again I shook the case over the hive and the small cluster fell into it. No more bees were congregating on the branch as I stood aside and watched.

"They are going into the hive now, so I guess I got the queen in that little cluster." I looked at Dave and we both laughed.

"That was way out! The most dramatic bee take I ever saw," he said. "I wouldn't have missed it for anything."

I left the hive partly uncovered until sunset, although we didn't think any scouts had gone out since we saw the bees arrive and took them immediately. But there were still bees in the air and some were lighting on the branch where the queen had last been. But it was too high for me to smoke as we usually did when we took bees. A little smoke puffed on the spot where the queen has been will remove her scent and bees will not go back there.

That weekend when Dave was well we took the hive out to the apiary, and now there were seven belonging to us.

Harvesting Honey

In the Dark Ages bees were killed to obtain their honey, but gradually such barbarous practices ceased and people took only what they thought the bees could spare. Nowadays beekeepers strive for a surplus of honey without endangering the bees, arranging to get their harvest from the summer-long honey flow. There are a number of honey flows each year; some last only a couple of weeks, since the fruit trees flow in the spring. Clover and alfalfa bloom in June and July, asters and goldenrod in August and September, and most wildflowers from June to September.

We set the seven supers in mid-June and in mid-September we harvested the honey, which was quite an undertaking for first-year beekeepers. We put on veils, gloves, and long-sleeved shirts, and tucked our jeans into our boots. We used the smoker to keep the bees calm, or scared as some beekeepers describe it, for this is when they can become really angry. Wouldn't you, if a giant lifted the roof from your house and began to take your food?

I lit the smoker and Dave used the hive tool to pry up a cover that had been sealed with propolis. I pumped the bellows of the smoker and puffed some smoke down into the super. Now we were ready to harvest our honey, usually forty pounds per super.

Another way to prepare a super for harvesting is to remove the super from the brood chamber, replace the queen excluder with an inner cover containing a bee escape, then replace the super for twenty-four hours and all the workers are out.

The inner cover is cleated on both sides, so there is a full bee space both above and below it. The bee escape is a hollow metal implement four inches long, one-and-three-fourths inches wide, and one-half inch thick. It has a spring beeway in it and the bees must squeeze out through a V-shaped opening that yields as they press against it. They can go out but cannot return. It is a one-way street for bees and clears a super in twenty-four hours. We did not use this method because our bees were out in the country: we would have had to go and place it the day before we harvested, and we could not well manage the extra trip.

Dave loosened a frame with the hive tool, pried up at both ends, and said, "Frame grip," which I handed to him. He drew the frame carefully from the super and said, "Bee brush."

I am assistant here, like a scrub nurse to Dave's surgeon role, and pass him the tools he needs.

"What lovely bees," I said as he held up the frame. "I hope the comb is filled." All I could see then was bees until Dave brushed them carefully back into the super, revealing a full comb of light-colored clover honey. Dave handed it to me and I put it into a hive-sized box and covered it immediately to keep bees of other hives from trying to rob an unprotected comb.

Dave removed four frames covered with bees, which he shook back into the super, then brushed off any remaining bees, and I stored them with the first one. The fifth frame Dave inspected and put back, saying, "This one needs a bit longer." Another one was put back as he said, "Not completely sealed." Two more frames were filled and two had to be left. He harvested two hives that day with five remaining to do as soon as we had the time.

Before we left, Dave took a quick look into the other hives so we would know what the yield would be.

"Four will have full supers." Then as he looked in the last one he whistled: "This one has no honey in the super, I wonder why. I'll have to look in the brood chamber later to see what is wrong. I'm going to take the super off now so we can use it to take honey home." He came out of the apiary carrying the super and we loaded the honey from the two colonies into the back seat of the car and went home. The box and super were set on newspapers and crosswise of the seat so they would ride smoothly.

"Did you heft the hive that had no honey when you took the super off?" I asked.

"Yes, and it seemed quite heavy, so they may have enough honey to carry them through the winter, in the brood chamber. There's still the fall honey to store."

The unproductive colony was one we had taken in June, but it was still going to see itself through the winter. It was also a small colony.

The work of opening hives, removing frames, and brushing off bees must be done quietly, swiftly, and carefully. Dave had discovered that if he shook each frame over the super, most of the bees fell off and there wasn't much brushing to do. This speeded operations considerably.

If robbing begins through carelessness or forgetfulness of the beekeeper, the work might just as well stop until robbing has subsided; otherwise the entire apiary will be in an uproar over the honey. I used an inner cover to top the super where we stored the harvested combs so there would be no robbing.

A robber cloth may also be used to cover a super of harvested combs. It is made of coarse cloth or gunnysacking about a yard square. Two laths as long as the super are laid one upon the other with one edge of the cloth between them and tacked securely. The cloth is longer than the laths, allowing six inches to hang from each end. The opposite side is treated the same way and the cloth is complete. To place a frame of comb in the super, take a lath in one hand, lift it from the super, quickly put the frame in, and cover it instantly. It is bee-tight.

For extracted honey the combs remain in the frames and the caps on the cells are removed with an electric uncapping knife in the extracting room. The frames are hung lengthwise in an extractor, the handle of which is turned manually or by electric power, and the honey rains out against the inside of the extractor like rain on a roof. After the frames are extracted on one side they are turned over and run again to extract the other side. The honey is drained from the extractor through a faucet at the bottom and through a sieve into a bucket or barrel for storage.

It was this extracting and handling of liquid honey that caused David to say he didn't want to get all stuck-up. But I am sure that producers of extracted honey manage it very well in a neat and careful way, as honey daubed on walls or floors with angry bees all around wanting it is very irritating.

After the frames are extracted they are left where bees can clean them up and then they are ready for reuse. Some beekeepers extract some combs from the brood chamber too, but we always leave the brood chamber honey for use of the bees in winter. Then we don't have to feed them.

Since we planned on selling comb honey, the frames were not left sticky. We simply cut the four-pound combs from the frames, which were then filled with foundation and ready for use.

Our Customers

"What will we do with over three hundred pounds of honey? That is what we will have when we harvest all of it," I said as we carried a super to the basement.

"Don't get uptight over it. I'll think of something," Dave said.

Dave did think of a way to sell our honey: he ran an ad in our southside paper; we had dozens of calls for comb honey, and our customers filed in. I was kept busy cutting the four-pound combs into one-pound pieces and wrapping them in waxed paper. If a customer bought eight pounds he or she preferred to have one-pound packages instead of two four-pound combs.

Most of our customers wished to take one teaspoonful every day of local comb honey to combat asthma, hay fever, or emphysema. They said they were benefited by it and told their friends. When they became aware that I harvested honey only once a year, in September, some of them took eleven or fifteen pounds to last the year.

They wondered how to keep it and I told them, "If you put it in the refrigerator it will granulate; all pure honey does. Store it wrapped as it is at room temperature of seventy degrees."

Many said that it didn't matter if it did crystallize, what was important was to have the comb. So by December we were out of honey except for what we needed for ourselves. I kept the names, addresses, and telephone numbers of our customers so I could notify them the next year when I had honey again. They were delighted with this.

I know that in my case anyway comb honey prevents hay fever, because I once suffered from it from spring to fall each year. But since keeping bees and using comb honey I am free of it. Comb honey contains pollen and that is what causes hay fever; when the pollen count in the air is high, people are miserable. Eating the pollen in a comb is something like having a shot for hay fever, only less bothersome and for me much more enjoyable.

Recently I read of a man in California who harvests pollen with a canister-type vacuum cleaner and sells it to drug companies who use it in making allergy drugs. He calls himself the Human Bee, and he will be busy as one until the pollen season is over in October. This would indicate that it is accurate that asthma, hay fever, and emphysema conditions are relieved if the sufferer eats comb honey. It has all that an allergy remedy could have plus the fact that is a fine food and easy to take.

From year to year we lost track of some of our customers but gained others who discovered the merits of eating comb honey.

Beekeepers and Beekeeping

Some beekeepers succeed and some don't, but anyone can keep bees if not timid and will follow the rules. It is best to begin on a small scale, as bee culture requires great skill. Fortitude is needed and a calm spirit, for you surely will be stung. The beginner will be stung more often than the experienced beekeeper, due to mishandling or irritating the bees. His body has not yet built a resistance, so the stings will be painful and swell badly, as mine did when Dave first opened a hive.

Anger and exasperation make stings a certainty; all movements should be slow and deliberate. Fear can bring stings also, as fear changes the body chemistry and bees dislike the result. Striking at bees is a sure way to be stung, as my father was. If a bee alights on you, let it crawl about unharmed and it will fly away, for bees never sting without provocation.

Furthermore, a beekeeper should be patient, for it can be two or three years before he becomes competent to handle bees without help. Beekeeping is not an easy way to augment one's income but it is an interesting avocation and it cultivates calmness of spirit, self-control, and patience besides supplying the luxury of honey. Done on a large scale it can support a family.

A woman especially needs an avocation, for the daily routine of housework wears the mind and soul full of ruts even if one loves it. Beekeeping frees the spirit from eternal drudgery and brings a great awareness of God's care for His small creatures. I always wonder what bees will do next and on the next visit I usually find out.

Hives are managed by females, so beekeeping should be as much a woman's work as embroidering, sewing, or painting—and it takes one outdoors. There is great happiness in putting foundation in frames, visiting the apiary, and cutting honey from the frames and into one-pound pieces to wrap in waxed paper to sell. Bees can be exasperating, but I don't react or they will also react and sting and the fun will vanish. Many people spend money on sports, games, or gardens with no idea of profit, so the pleasure of beekeeping cannot be seen in the profit and loss columns. Two or three colonies should pay for themselves, plus honey for the owner.

All this logic is called for because a beekeeper is sometimes considered to be a bit odd, especially if it is a woman who keeps the bees. Occasionally when Dave and I have answered a call to pick up bees, someone has commented: "A woman beekeeper! Imagine that! I always thought women were afraid of bees."

I have no difficulty imagining a woman beekeeper because I am one. The only difficulty, as I see it, is that some of the work is a bit heavy for a woman, since supers of honey weigh at least forty pounds. When I harvest honey alone, I solve this by dividing the frames into three sets, making the weight I carry at any one time not over fourteen pounds.

I remove a frame of comb from the super, clear the bees, and put the frame into an empty super that I keep nearby. When I have three frames in it I carry it to the car and put the frames in another empty super, go back and remove three more, and so on until I have all of the honey. One of my bee manuals states that harvesting honey requires two men, but I have been doing it alone since Dave went away to college. He is a senior now and says that he will keep bees again after he graduates.

I always wear the bee veil as I work, because the face is a favorite place for a bee to sting and the reaction is greater there. A bee intent on stinging has a louder than normal buzz which attracts other bees. The smell of the venom also attracts them and they come to sting the same place. If I am stung, even on my clothes, I puff some smoke on it to kill the odor.

Stings are what make beekeeping a select calling. Everyone would have a colony in his garden if bees were stingless. The sting and venom sac remain on the skin when a bee stings, as the sting penetrates about one-twelfth of an inch. It should be flicked off (not picked) at once as the muscles still work to push the barb deeper into the skin and pump in all the venom. The bee dies in a few hours, if that is any consolation.

A bee-sting feels like a sharp pinprick at the time and it may swell and itch for days. After the sting has been flicked off, cold water is the best remedy, as it dilutes the venom and checks inflammation. One should never squeeze the sting out—that squeezes more venom into the wound—nor should the mouth be applied, as the venom is harmful to the digestive organs.

If only a few of the countless remedies for bee-stings were effective, there would be no reason to dread being stung. If the sting is pushed out immediately it will rarely produce serious consequences. The wound should not be rubbed, as this diffuses the poison over a large area and pain and swelling will follow. If you are perspiring, stings are less painful as some of the poison is exuded with the sweat. In severe cases, ammonia has been given in quantities of from five to twenty drops in hot tea. This causes perspiration and neutralizes the effects of the poison. The venom will produce less and less effect upon the system and beekeepers of long duration acquire an immunity and even seem to thrive on stings. An elderly, bald beekeeper had stings covering his head but had no unpleasant effects from them. If a beginner allows himself to be stung frequently he will become accustomed to the venom in two seasons. He will become inoculated with bee venom just as sure as vaccine prevents smallpox. Sometimes I am stung through my shirt or jeans, but the sting and most of the venom remain in the cloth, and later I can scarcely locate the wound.

Some doctors maintain that bee venom is beneficial for rheumatism and arthritis because it contains cortisone (which is never given until the symptoms are rather severe). A few stings have caused the disappearance of these afflictions in the practice of a certain doctor. He keeps a hive of bees outside an office window, which has a sliding panel to give him access.

When an arthritic patient comes, he slides the panel back, takes a bee with a tweezers and holds her against the shoulder of the patient until she stings. The patient gets the venom, with cortisone, in the troublesome joint.

I wear white jeans, shirt, and sweater of late, and all is very peaceful. Formerly I wore dark jeans and a plaid shirt and bees were crawling all over me, some stinging as I worked, yet there were none on my white sweater. At last I got the message, bees dislike dark colors and are more likely to try to sting through them. One-piece white suits are also available; they zip up the front and have elastic at neck, cuffs, and ankles. If one feels clumsy with bee gloves he may wear a lighter rubber glove or work with bare hands provided he has rinsed them in a solution of carbolic acid, three ounces to a quart of water. This will prevent bees from alighting on the skin and stinging. Venom has a sharp, pungent odor which angers bees, so a person once stung is a further target.

Mr. L. L. Langstroth was America's first noted keeper of bees. He was born in Philadelphia in 1810 and is called the Father of American Apiculture. He was interested in insects from early youth and began to keep bees in 1831, after seeing bees in a friend's attic. He bought two colonies in old box hives and devised the movable frame hives used in preference to all others today.

Some beekeepers truck their colonies south when the local honey flow is over, as there is no cessation of growth due to frost there. Others are content to garner the harvest obtainable in the north. Many rent colonies to orchard owners in the spring to get the fruit honeys as well as to cross-pollinate for a good yield. Although it is widely known among orchard owners that the presence of bees increases the crop, there are still some who think bees injure fruit, and a man who brings in or owns bees is considered a bad neighbor.

Bees do not feed on fruit: their jaws, adapted to manipulate wax, are too feeble to puncture the skins of fruit. An orchard owner checked his trees and found many bees banqueting on bruised or damaged fruit and even at the moist stems from which grapes had recently been plucked, yet no solid fruit was harmed. To further prove his point he cut off a sound cluster of grapes, laid it on a table and pricked one grape with a pin. The bunch was soon covered with bees, but at sunset when the bees left, only the punctured grape was depressed, while the bloom was off the bunch and the grapes were shiny.

Another test was to place a bunch of solid grapes in each of five hives, directly over the frames and protected by an empty super. After three weeks the bees had glued them fast with propolis, but not a grape was damaged. The main damage to fruit is done by birds, then bees take over and receive the blame. Wasps and hornets also damage fruit, as they have strong sawlike jaws that can penetrate even the toughest fruit skins.

Michigan has a thriving "rent-a-bee" business; bees are in such demand for pollinating purposes that there are shortages in some areas. Orchard owners rent colonies for ten dollars apiece in many parts of the United States; in fact, the pollination of crops is worth far more than the honeybees produce. Man has changed from a robber to a protector because bees are valuable; thus bees now exist in much greater numbers than nature could provide. The skillful gatherer from wild bees once could harvest one hundred pounds each year with hard work and pain. Today a commercial beekeeper harvests tons of honey from a thousand or more colonies under his care.

Bee Products

The product that everyone is most aware of is honey, but there are other less known but important products such as comb, pollen and especially wax and propolis.

History doesn't record the discovery of honey, but all peoples of the world have been using it for millennia. The Bible mentions honey a number of times: Samson ate honey that he found stored in the carcass of a lion, Jacob sent a gift of honey to a friend. Canaan was a land flowing with milk and honey, John the Baptist lived on locusts and wild honey, and Jesus ate fish and honeycomb after His resurrection.

Archaeologists say that Neolithic man ate honey—and bee larvae and eggs as well. Egyptians used honey: they kept bees on barges on the Nile and floated them along as the nectar areas changed.

For centuries honey was the only sweet in most of the world except China, where sugar cane seems to have originated. At one time tribute was paid in honey, as were taxes. Greek mythology states that honey was the food of the gods on Mount Olympus; Roman women mixed honey and milk as a lotion for their skins.

The Jews knew that honey was a vegetable product, but since

it was produced by bees, which were unclean, they asked a rabbi why they could eat it. The rabbi's answer was: "Bees do not make honey, they only gather it from flowers and store it in their homes."

But the rabbi didn't know all the details. Bees do gather nectar from flowers but then it goes to their honey stomachs where it undergoes a chemical change before it is stored in the combs as honey. It is still quite thin and is allowed to evaporate in open cells while the bees fan the air with their wings to aid the process, forcing a strong current of air through the hive. When the cell is full the bees seal it with a thin cap of wax.

The cap is begun at the bottom of the cell and raised as nectar is deposited until the cell is full. Nectar varies in quantity in flowers, depending on the time of day and the weather. It is most plentiful in the morning, decreases as the sun gets higher, and after three in the afternoon again increases until after dark. Rain dilutes nectar in the flowers.

Honey is a food and is predigested by the bees. It is used on cereal, fruit, biscuits, toast, and waffles as well as being eaten by the spoonful. It is used in the making of cakes, candies, cookies, jellies, vinegar, and many other foods.

When used in baking, one-fifth cup of liquid should be omitted from the recipe for each cup of honey used; one-third cup of honey equals one cup of sugar. Two tablespoons of honey, four tablespoons of lemon juice, and three tablespoons of oil make a nice salad dressing. If the oil is omitted and one-half cup of sour cream is added, the dressing will be excellent for fruit.

Early doctors used honey as an antiseptic: it was mixed with flour and used to cover boils and draw them to a head, or on burns to keep out the air and promote healing. Mother put a thinly sliced onion in a cup, covered it with honey, and let it stand on the back of our coal stove for several hours. We were given this liquid for coughs, colds, and sore throats; it was very effective and we enjoyed taking it.

Honey is not stored in the worker cells but in the larger drone or storage cells, which are about the diameter of a lead pencil and three-fourths of an inch deep. Honey from different flowers is vastly different. Clover honey is very light in color and most in demand, alfalfa is next, and the fruit tree honeys are also in demand. Bees must bring in three to four pounds of nectar to make one pound of honey. Honey is sterile; yeasts, bacteria, or molds falling into it die or remain dormant.

Bees store honey at the rear of the brood area, above it and as near to it as possible. When the brood chamber is full the bees store it in the super and the beekeeper gets his harvest. The quality of the honey depends on the source of nectar, not on the secretions of the bees. Apple blossom, white clover, alfalfa, buckwheat, and other varieties of honey each have their own flavor and color. The difference is so great that people, tasting the difference, might imagine adulteration. The lightest in color is not necessarily the best; some pale yellow honeys are very fine in flavor. Buckwheat is very dark but excellent, and honey from most fall flowers is dark and has a stronger flavor.

The honeycomb is one of the wonders of the insect world. It consists of myriads of hexagonal cells one-half inch deep and tilted slightly upward. There are twenty-five to the square inch on each side of the comb. Bees are master architects and produce their own materials. To make wax for comb building, young bees that have not yet gone to the fields gorge on honey, lock their claws together, and hang in a living chain or curtain from the top of a frame to the bottom. Their body temperatures rise as they digest the honey and in several hours wax flows from pockets on both sides of their abdomens in flakes like fish scales. They scrape these off with their hind legs and other bees take and chew them into caps for brood chamber or super cells or for the cells themselves. This building of the comb is the result of a moving mass of bees, no one bee remaining at work on the same spot for more than a minute or two.

Each bee contributes something to the construction of the comb, but no single bee completes anything; yet the efficiency in building comb is one of the wonders of God's creation. One square foot of comb is made from three ounces of honey and will hold ninety ounces of stored honey or six thousand baby bees.

In the natural state bees attach comb to the roof and sides of the recess where they live, but a beekeeper gives them hives with movable frames containing sheets of wax foundation. This foundation is a thin sheet of beeswax embossed with the hexagonal pattern of comb cells and the bees draw this out and make the comb. Bees do not really need this pattern as it was copied from comb, but in the frames it keeps the combs even and straight and prevents them from collapsing from the heat on a summer day.

Foundation for the super is very lightweight but in the brood chamber it is laced with wire for extra strength, and becomes as dark as chocolate from many seasons of brood being raised in it. As brood hatches the pupa shells remain in the cells, and although bees clean the cells, they become dark by the end of one season.

Brace comb is built in every hive and super: small spurs of solid comb built between the frames to prevent vibration. It never contains any honey, and makes removal of the frames difficult. It must be cut with the hive tool before a frame can be lifted out. Frames hold the combs apart besides keeping them straight, and furnish a five-sixteenth-inch space where the bees can travel and work without interfering with one another.

A comb consists mostly of hexagonal cells because this gives the greatest capacity and strength in the least space and requires the least labor. However, there are four-sided cells at the base and even some triangular in odd corners, so that all space is utilized. The hexagonal cell is produced naturally by the bees, as a cell wall contacts the next one it flattens and becomes angular.

Combs are about an inch thick with cells on both sides and the distance between combs is half an inch, so the bees can work comfortably. New combs are white when first built but gain color when yellow nectar is stored; cells are not uniform in size. Drone cells are larger, with eighteen to the square inch, compared to twenty-five to twenty-seven for worker cells. Pollen must be available, since comb cannot be produced without it.

A newly hived swarm always begins construction on worker cells; with a prolific queen, very few drone cells are made. Should a queen die or be removed during comb building, all the cells will be large. Queens prefer to lay in small cells; but, strangely, workers prefer to build large cells.

In a newly occupied hive where there are as yet no cells, the queen follows the builders closely and no large cells are made, so she lays the brood. After three weeks, the bees from the first-laid eggs hatch and leave the cells, which the queen returns to fill again. The workers, free from restraint, follow their preference and large cells are made at the back of the hive.

Should the honey flow cease, comb building and laying also cease. When another honey flow starts, building is resumed, but the queen is not then among the builders waiting for cells. She is in the other end of the hive laying in the newly emptied cells, so storage cells are again made, as much as a third of a frame.

If the weather is extremely hot and the hives are in the direct rays of the sun, bees have a powerful ventilation system by fanning their wings to keep the combs cool. If, however, the heat is intense enough to drive the bees outside, heavily laden combs may melt down, causing honey to drip from the comb. If one comb melts, others may fall also, crushing brood, queen, and any bees remaining in the hive.

This is largely avoided when frames are used, but even then combs may break loose from the top and collapse. The solution is to have at least part-time shade for the bees.

Few things impress the mind as greatly as the truly scientific way the ventilation of the hive is done. By stationing themselves among the combs and along the entrance, then fanning with their wings, the bees create a flow of air through the hive that keeps it at the proper temperature most of the time.

The advent of foundation marked an important step in bee culture, as it made straight comb possible. One frame slightly out of line is sufficient incentive for the bees to fasten two combs together. If a comb breaks loose at the top, the beekeeper will find at least two combs welded together.

Pollen, the fertilizing dust of flowers, is gathered by bees from blossoms and is used in feeding larvae; in fact, brood cannot be raised without pollen. Without honey, however, the mature bee cannot survive on pollen.

Gendelach, a German apiarist, confined a colony with a fertile queen in a hive supplied with honey but no pollen. Comb was rapidly built and the cells filled with eggs which hatched, but the larvae died within twenty-four hours from lack of pollen.

Fresh pollen is gathered by bees even when there is a supply of old pollen in the cells. When nectar and pollen can be obtained from the same blossoms the industrious bees gather loads of both. They store the pollen away by backing into cells and brushing it from their legs. It is then packed down by hive bees, covered with honey, and sealed with wax, since it is easily decomposed by mold.

Pollen is usually stored in worker cells and varies from yellow to red to brown, depending on the plant from which it came. Only one kind is gathered for a load. Bees gather and use as much as one hundred pounds of pollen in a year, and if they can't find pollen early in the spring they will take in flour, meal, or sawdust as a substitute. It is a common practice for beekeepers to supply these in this season in shallow boxes, filled about two inches deep with meal, rye, or finely ground oatmeal and set near the hives. These should be firmly pressed down with the hands so bees will not drown in them. The boxes are baited with old combs or a little honey to attract bees. Bees prefer these flours to old pollen in the hives.

This is how bees can be bred early in the spring, which is desirable if the apiarist intends to trap the swarms to increase his colonies.

Propolis, the heavy bee glue obtained from resinous buds and the shiny leaves of plants and trees, is used to coat the inside of the hive, closing all crevices to keep out wind and rain and to cement the frames down. When gathered it is bright yellow and so sticky that bees never store it in cells but apply it at once. It adheres very firmly to the bees' legs and is carried like pollen. A bee scrapes it from the buds and leaves with her mandibles and transfers it to the pollen baskets on her hind legs. When she returns to the hive it is removed by hive bees who tear it away and shred it with their jaws for immediate use.

A mixture of wax and propolis is stronger than wax alone and strengthens the attachment of comb to frame, so there is no vibration in the hive. Summer heat keeps it soft and pliable. The extensive use of propolis is why an established colony can be moved easily without angering the bees. It gets on the fingers of hands or gloves when the apiarist handles the frames and is removed with turpentine, alcohol, or ether. It can be scraped from wooden surfaces with a knife or the hive tool.

Propolis has been used by bees to form a border around the edge of the shell of a snail that had crawled into the hive, thus anchoring it to the floor. A shell-less snail was first stung to death, then covered with propolis to prevent decay. A mouse, stung to death, then liberally coated with propolis, was in one corner of one of my hives last spring. It resembled a small piece of gray wood more than a mouse and I wondered if a mummified mouse would look like that.

Uses have been found for this unusual glue. Dissolved in alcohol and filtered, it is used as a varnish and gives a high polish to wood. Incense to perfume rooms and halls is made from propolis and a number of other ingredients. In Russia, wooden tableware is made water-repellent by coating it with a mixture of propolis, linseed oil, and beeswax.

Beeswax is heavier than some other waxes and melts at 144° to 150°. It is used to make foundation for hives and supers. For hives it is made as thick as brown paper and reinforced with wire to keep it from breaking under the constant weight of brood. In the super foundation it must be very lightweight, as light as the finest machines can make it to avoid the "fishbone" central rib in comb honey caused by too heavy foundation in the frames. The bees thin out the foundation and draw cells from it.

Beeswax is used in many ways. Perhaps its earliest use was before the invention of papyrus or parchment—and even after, as it was inexpensive. Boards were covered with a light coat of beeswax and written on with a wooden pen called a stylus, sharp at one end to write and flat at the other, to erase.

In ancient times, after it was discovered that beeswax did not decay, the Egyptians used it in embalming their dead. Today it is used to make candles, sculptors use it to varnish their work and to make figurines, and artists use it in batik by melting it and adding turpentine and color.

Furniture polish is made by melting one-half pound of beeswax, stirring in a cup of turpentine, adding one-half cup of alcohol, and stirring to a paste. A small amount is applied to the furniture and rubbed thoroughly. One of the newest household polishes to appear on the market contains beeswax. In Europe the floors and stairs are rubbed with beeswax, then carefully scrubbed with a dry brush every day until they shine. This eliminates the need for rugs.

Beeswax is also used in creams, pomades, ointments, and salves. We furnished wax to the art department of a local high school. To clarify the wax for use, we melted it in water with one tablespoon of vinegar to the pint. The dross on the bottom of the cake could be scraped off, but a better way, we found, was to put the wax in a cloth with some rocks to weight it down, then melt it in the water. This gave us a clean golden cake with the dross left in the cloth.

Wax is also used in varnish and in incense along with charcoal, and on woodware to waterproof it. Most of the wax in commercial use is furnished by the comb cappings from beekeepers who produce extracted honey, as they reuse the frames of comb. We sold the wax comb with the honey as comb honey, which incidentally can't be adulterated or duplicated. For over fifty years there has been a standing reward of a thousand dollars for anyone who can produce honeycomb, but it has never been claimed.

A Burning Event

In March of the year that all of our bees were out in the country, Dave took a call that changed our plans for the year. His side of the conversation was as follows.

"Gosh! Who did it? . . . Why would he do a thing like that? Did you lose all your equipment? . . . Are ours burned too? . . . Is there anything we can do? . . . We sure will . . . Yes, I'll keep in touch."

He turned to me with a look of unbelief on his face. I asked, "What was that all about?"

"What a bummer! Someone threw gasoline on Mr. Johns's bees and burned them."

"When did that happen?" I asked in unbelief.

"Just last night—what a deal!"

"Are ours burned too?" I asked.

"Luckily, no. They are sort of off by themselves, you know."

"Does Johns know who did it?"

"He's not sure, but he thinks it was the man who lives across the lot in that little white house. Of course he can't prove it, but that man was always mad, and he once said he hated the bees there. Well, there's nothing to do now but clean up the mess. I wish I had time to help him, but midsemester exams are on and I don't. Too bad."

Burned hives

"I'm surprised that the man objected to the bees being there. They must have been at least one hundred feet from his house," I said, "and I don't think they ever bothered him."

"That's what Johns said, and the ground between belongs to the friend who told Johns he could keep his bees there. Oh yes, he said we better move ours to another place where he has some."

"We better do it this weekend. Will you have time?"

"I'll take time. Lets go out late Friday and stuff the entrances, then we can move them Saturday," Dave suggested.

We moved them to a very nice place that had an irrigation ditch running close by and many big trees for shade. There was always water in the ditch, so we were relieved of having to supply the bees. We took two hives at a time in the trunk of the car, and as we placed them we leaned boards up over the entrances, pulled the stuffing, and the bees were in business again after taking account of their new location.

On the way home Dave said, "Johns asked if we could take

bees for him this year. That's how he said we could help. OK?"

"Of course. We have seven colonies and they may let out swarms. And I'll not have to make hives and supers," I said with relief.

Strangely enough we never got a swarm out of our bees while they were there. I'm sure some came out but we were not around to take them.

"He will pay me for the swarms," Dave said.

"How much?"

"He didn't say, but I imagine around five dollars a swarm. He would have to pay twelve dollars for a small swarm if he bought them and two dollars more for a queen. The swarms we take will be good-sized, about twenty thousand bees with a queen that is already acclimated as they are."

"That will be a nice little sum, but somehow I think you should have had a definite understanding with him about the price."

So that year we took bees for Johns, nine swarms, which he picked up of an evening when Dave told him we had a swarm. He had brought a number of hives for us to use, as we had no extra ones. Dave was delighted when Johns came for the last one, anticipating his pay. Johns counted out nine one-dollar bills and handed them to Dave, saying very jovially, "Thanks a lot, Dave. You are a good scout," and he slapped Dave on the shoulder.

Dave took the bills silently and the amazement that spread over his face was unnoticed by Johns. We watched silently as he left, then Dave said, "Golly! I don't dig it. Nine dollars instead of forty-five, he sure put me on."

"But you never had a definite arrangement and he did call you a good scout," I said. "I guess sympathy comes cheap."

"Never again," Dave said and fell silent. Then he added, "Maybe he thought we owed him something for a place for our bees."

"Could be," I said. "We can be glad he had to bring his own hives. Now I will have a long time to make hives for next year, if you still want to take bees."

"I sure do," and Dave was all smiles. He pocketed the bills thoughtfully. "I sure blew it with Johns. Oh well, live and learn."

"We will have plenty of honey this fall, the bees will do well out there, and we have supers from last year," I said.

We set the supers and by fall had three hundred pounds of honey and our list of people waiting to buy it. We had over $250 to divide between us, since our only expense for the season had been for super foundation, which was less than ten dollars.

Bees at School

In May of the following year Dave called me from junior high school, where he was a student. It was just after twelve o'clock and he was very excited.

"There are bees on the hedge in front of the school and the principal wants to know if I can take them. I'm calling from her office."

"What do you want me to do?" I asked.

"Could you put a hive and five frames in the car along with my gloves, veil, and bee brush and come over?" This left him breathless and I knew the principal must be listening.

"When do you want me to be there?"

"As soon as you can make it. I'll be excused from class to do it."

"Where will you be?"

"In class, so you come to the office and ask for me."

"I'll be there in about twenty minutes, OK?

"Good! I'll be waiting for the office to send for me. Bye," and he hung up.

I went to the office, identified myself, and asked for Dave. A messenger was sent to his classroom and in a few minutes he came in. Together we went to work to put the bees in the hive.

It was not difficult, as the hedge was a low ornamental one about three feet high with just enough room beneath the branch to set a hive. Dave put on his veil and gloves and I laid some newspapers on the ground under the hedge. Dave shook the branch on which the bees were clumped and there was a shower of bees into the hive and on the paper. We waited for a while until we saw the bees were going into the hive and we knew we had the queen. Dave put the cover on at an angle so the flying bees could go in.

"Do you know when they came?" I asked Dave.

"At second lunch hour, just before I called you."

"I don't suppose there are any scouts out that soon, so maybe I should take them home now," I said.

"I think we should wait until school is out," Dave answered.

"You don't think any of the kids will bother them?"

"No, I think they would be afraid of them, that they might get stung."

"Why don't we ask the principal?" I suggested.

The principal assured us that the hive would be quite safe under the hedge until school was out, so Dave went to class and I returned home. But the principal hadn't reckoned with the host of junior high boys, for when I returned at three fifteen the hive had been turned over and bees were all around. A couple of frames were lying on the ground with holes punched in the foundation and the cover was several feet away on the grass.

I waited nearby and Dave came out in a few minutes. He was very indignant at the lack of consideration shown by some of the students.

"Such a bunch of idiots! Why couldn't they have left this alone?"

"Oh well, it is done now, so let's see if we can get things put together again. I guess the wise thing would have been for me to take it this noon."

Dave put the hive back on the bottom board and inspected the frames that remained in it to try to locate the queen.

"Cool! Here she is," Dave said as he looked at the third frame. "Isn't it lucky that she didn't run out on the grass where we couldn't find her? Now the bees should go back soon."

A queen is very shy and will remain with any fair-sized group of bees, especailly when on a frame with foundation, which is very fragrant and must give her the feeling of being in her former home.

As we waited for the bees to return to the hive, streams of students swirled around us, most of them surprised and many curious and interested.

"What goes on here?" "Oh, my gosh, bees!" "Where did they come from?" These and more were the comments and questions, which Dave answered when he could. They were awed to learn that we were beekeepers and would take the bees home.

Dave said, "Some goon knocked the hive over and the bees were spilled out. It's just too bad he didn't get stung."

"Maybe he did," a girl said.

"Not a chance, bees that are swarming don't sting. I put them back just this way," and he indicated his shirt sleeves and no hat.

By four o'clock the students were gone and the bees were in the hive. Dave tied a rope around it so it would travel safely, stuffed the entrance, and loaded it into the trunk. We drove home.

"How did the principal know about you and bees?" I asked.

"We were talking one day at lunch and I told her."

Dave put them in the backyard and the next time we went to the apiary we took them along.

On the Pickup List Again

I had made hives and supers during the winter and we were ready to pick up bees again by May.

"Any bees to pick up today?" Dave would ask as he came home after school.

Most of the time I would have to say "None today." But this day I had better news: "There is a swarm on a small spruce tree in a shopping center on Colorado Boulevard."

"Great! Let's get going, I will study later."

"It's a paint store. When the man called he said he wanted us to be careful not to break any branches."

Dave laughed, "Pretty particular, isn't he? Most people want to be rid of the bees at all costs."

We got together our usual equipment, hive body with cover and bottom board, five frames with foundation, bee brush, newspapers, rags, and the smoker. We wore gloves and veils as we had planned to do, just in case there were undernourished bees. As we drove up and began to unload our equipment, a man came out of the store and looked us over.

"Well, I never! A woman beekeeper!"

"And a good one," I laughed.

"I see you're all prepared," he said as he rubbed his hands together nervously.

From his attitude it seemed he had not seen beekeepers before.

"Now if you'll just not harm the tree, everything will be fine."

Dave had been looking the situation over. "The bees are mostly on this one branch which is heavily weighted down." Then he turned to the man and said, "We will not harm the tree at all, but you better go inside the screen for safety."

Dave put the hive on a tall stool, which the man gave us, under the branch after I had spread newspapers on the ground as far under the branches as I could. This was so we could see the queen if she fell outside the hive when Dave shook the branch. I sat nearby to watch for her.

Dave took hold of the branch, carefully raised it to as near horizontal as he could, and shook it over the hive. A great shower of bees fell among the frames in the hive, with very few on the newspapers and no queen among them.

"I guess that did it," he said.

The man came out smiling. "That was the neatest thing I ever saw. I thought you would be working here until closing time," he was tall, thin, and gray, and had a very attractive smile.

"When did the bees come?" I asked.

"Just this afternoon, shortly before I called you."

"In that case we can take them with us when we go, because they haven't had time to send out scouts," I said.

"What do you mean by scouts?" he asked.

"Scouts leave the cluster to try to find a place where the swarm can live," Dave told him. "Sometimes that takes several days. But we have solved their problem."

"Then it is a good thing we saw them come and called you right away, isn't it?"

"Right," Dave said.

As we watched, bees were flying from the ground to the hive, so Dave lit the smoker and puffed some smoke on the branch where they had been so the flying bees wouldn't alight there.

The man had gone into the store while we finished our work, but he came out with a ten-dollar bill which he handed to Dave, saying, "We want you to have this for doing such a fine job."

"Thank you, but the swarm of bees is our pay for taking

them. We will take them to our apiary with the others," Dave said.

"Oh, so you are going to keep them. I thought you would dispose of them."

"Oh, no. We are beekeepers and want bees."

"For goodness' sakes! I thought you were just someone the fire department had sent to get the bees and destroy them." He reached over and tucked the bill into Dave's shirt pocket. "Now you leave that there, I want you to have it." He took each of us by the arm and led us into the store.

"I want all of you to meet a couple of very unusual people," he said to a group of employees in the office. The salesmen flocked around and even the customers in the store and we had to tell them about our work with bees. It was a very abbreviated story with only the necessary details and when it was told someone said, "So you are going to keep our bees. When will they have some honey?"

"Most likely by September," Dave said.

It was May. He knew there would be honey by fall, as it was a very large swarm of at least thirty thousand bees.

"Won't that be fine. We will have honey from our own bees!" another employee said excitedly. Then she asked, "How will it come? In jars?"

I presumed she thought it would be strained honey, but could scarcely keep from laughing as it sounded so much like she thought we put jars in the hive and the bees obligingly filled them.

Dave said, "Most likely it will be comb honey. We don't extract because most of our customers want honey in the comb."

"Well anyway, we all want some, don't we?" and she looked around at the others. "Should we order now?"

"No, that won't be necessary. We will stop in September when we harvest the honey and you can have as much as you want," I said.

They were all very happy people and wished us well as we left to remove the hive. There was no activity when we got back, so Dave tied the hive securely, stuffed the entrance, loaded it in the truck, and we drove home. The man came out to get his stool and waved goodbye.

Dave leaned out of the car to say, "If you see any bees on the branch in the morning, just call us and we will come and get them. However, if there are only a few they will return to the colony from which they came."

We were not called, so they had no further bee problem. In the fall we stopped by with a large number of wax-wrapped pounds and everyone took some. They insisted on paying us even though we had not intended that they should. We liked them very much because of their interest in bees and us.

"Bet your life we pay," said the man who had talked to us in May, evidently the manager. "How is a fellow to make money if people don't pay?" and he patted Dave on the shoulder.

For several years we stopped there in the fall with honey, until at last no one was there who had been there when we took the bees, not even the manager. We hoped they were happy wherever they were, for they were a group of lovely people.

Bees in a Car

One day a call came from a woman in a shopping center saying that there were bees inside her car.

"I went into the store and left the window of the car down and when I came out there were bees all over the dashboard and I was afraid to get in. Can you come and get them?"

"Where is the car?" I asked.

"In the Athmar Shopping Center by the supermarket. It is a light blue car with the windows down and you can see the bees. You can't miss it and I live up the hill if you need me." She gave me her address.

We took our usual supplies and drove over, but here was where we met our Waterloo through no fault of ours. Dave put a hive on the floor of the car under the dashboard, swept bees off into a frame, and put it in the hive. He brushed all of the bees onto frames but they would not stay in the hive.

"That means that we don't have the queen," he said at last. "I wonder where she is, she must have split."

"She must have crawled into a hole somewhere in the front of the car. Do you suppose she is under the hood?" I asked.

Dave lifted the hood but there wasn't a bee in sight, yet we heard them buzzing somewhere. "They must be down in the body where we can't reach them."

There were still bees in the car on the dashboard but none in the hive. We watched and saw some slowly going into a small opening. "They must be going into a small place where the queen went, because most of them can't get in, but she isn't going to come out," I commented.

In the meantime a reporter from the *Post* had come to get a story on the bees—we never discovered who sent him. He took pictures of the car, the bees, and incidentally of us. We did not intentionally pose but he did get a shot of us across the car.

"Don't give our names," Dave whispered to me.

Just then the reporter came around the car and asked, "Would you give me your names for the story?"

"I prefer to remain incognito," Dave said. "We have more bees than we need and don't want any publicity."

The reporter left, and the next day there was an interest story in the paper with a picture of the car, the bees, and us intent on our work.

At length we decided we were not going to be able to take the bees and Dave went to the woman's house and explained the situation. She was very understanding, but I had strolled over as Dave seemed perplexed.

"What should I do since you can't get the swarm?" she asked.

"The only thing I can suggest is that you call an exterminator and have him kill the bees. They may not come out of the car at all, or they may in several days if they find they are too crowded," Dave told her.

"Can you do it?" she asked.

"No, ma'am, we are not prepared. We are beekeepers and do not have anything to kill bees. I'm sorry we could not take them but the queen went into a small opening, cars have so many of them. I think she had already found the hole before we arrived."

"Oh, it isn't your fault, I shouldn't have been so stupid as to leave the window open. I sure have learned my lesson, but wait a minute, will you please?"

She went into the house and came back with two dollars which she handed to Dave.

"No thank you," he said, "we never charge for taking bees, the swarm is our pay, but we just couldn't get this one."

"That's why I want you to take this." She reached for his hand and put the money into it. "That's something for all your work out there in the sun."

Dave smiled, said, "Thank you," and came down the steps. The woman came over to where I stood and said, "You don't mind, do you? About the money, I mean. You have a very fine young man there."

"Of course not. It was very nice of you, he felt quite bad about it. Goodbye," I joined Dave at the car where he was putting our equipment in the trunk and we returned home, two very frustrated beekeepers.

A Cluster Over a Garage

Another surprising bee pickup was for some bees clustered on a tree branch hanging low over a garage. We drove to the house about four o'clock in the afternoon, and the people, plus some neighbors, were in the backyard watching the bees and talking.

Dave strolled over and asked, "You have some bees that are troubling you?"

"You bet we have," the man said and he pointed to the big cluster up in the tree. "They came as I was going to work this morning. It's a good thing they are up there away from the kids, so no one got stung."

"We called the fire department but they haven't come yet," a woman, evidently his wife, said.

"They have now, we're from the fire department," Dave said. "Do you have a ladder?"

"Yes, there it is leaning against the front of the garage."

"We'll need it to get up on the garage and take them," Dave said as he started for the ladder.

Then the surprise came: "Oh, no you don't! You don't touch them! The fire department is going to come and get them," the man said angrily.

Suddenly Dave realized the man had not listened to what he

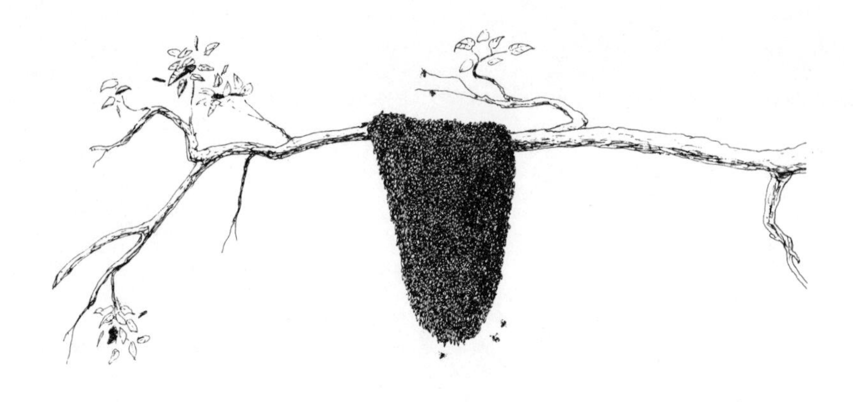

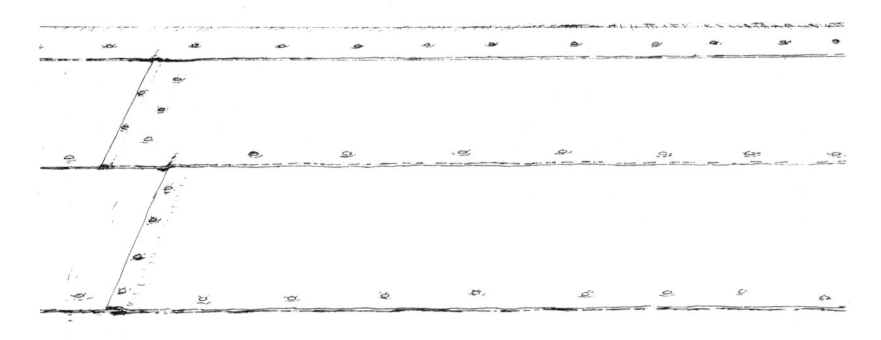

Swarm clustered over garage

had said, so he repeated, "We are from the fire department."

The man looked at us incredulously, a tall slender teenager and a slight, older woman, and said, "I don't believe it!"

"I hope you're not expecting a burly fireman with a hook and ladder car. We represent the fire department on bee pickups. We are beekeepers and are on the fire department's call list," Dave said. "The number they gave you was our number and I talked to you."

The man scratched his head and thought. "Come to think of it, they did give me a number."

Dave repeated our number. "And I talked to you."

The man shook his head. "I can't remember the number."

But he would not believe we were the right people, so before he decided to order us off the place Dave asked, "Can I use your telephone? I'll call the fire department and you can talk to them."

That turned the trick. We all went into the house and Dave dialed the fire department and said to the desk sergeant, "This is Wells, on a bee pickup, and the people here want confirmation that you sent me."

He handed the phone to the man, who listened intently, flushed, and hung up. He turned to us and said, "I'm sorry I didn't take your word about it, that fireman was tough about it."

"That's OK. Maybe we should have credentials of some sort but we don't," Dave said.

I wondered if even credentials would have done the work: the man seemed to want the excitement of a hook and ladder car.

The bees were difficult to take even with the ladder, for the garage roof sloped like that of a house. Dave finally got up and sat straddling the ridge; he looked down at me, grinned, and waved. Then he tied a rope to the tree limb and let it down to me so I could send up his equipment. I got the hive, frames, bottom board, and bee brush and struggled up the ladder with them. I tied the rope around the hive and its bottom board, put the bee brush in it, and Dave pulled it carefully up the roof to the ridge.

"The hive will slide down the roof. Can you get something to brace it?"

I rummaged in the car trunk and found another rope, which I held up. He looked surprised, then said, "Good. Bring it up."

I climbed the ladder and tossed it to him. He made a double swing of it and swung the hive on its bottom board, bracketing the cluster of bees.

"I'll bet these bees will be the first to have a swing of their own," he said, laughing.

I had to laugh too and the man, who was watching, simply roared. "That's way out!" he said and slapped his wife on the shoulder.

Dave had to brush the bees onto the frames and put them in the hive himself, since I couldn't get a foothold on the roof and the branch was too large to shake. But it was more handy than usual, for the hive was usually on the ground and he was on a ladder, now it was just below the bees.

When all the bees were in the hive Dave came down and told the man, "We'll leave them swinging there until after sunset, so we get all the scouts."

The man patted Dave on the shoulder, "So now you're telling me bees have scouts. OK, I'll believe it. I never saw anything so well done, you're really something." He was still laughing as we drove away.

When we came back the ladder was still in place; the man got a flashlight that he turned on the hive. Dave went to the top of the roof with a hive cover, his hive tool and a length of cloth. He secured the cover in the rope that already held the hive, stuffed the cloth in the entrance, and was ready to lower the hive.

While he was doing this I asked, 'Do you have another ladder? I will have to go up to the roof to get the hive and Dave will have to have a ladder to come down."

"My neighbor has one," he said, and got it and put it up against the garage.

Dave lowered the hive slowly in its swing until I could reach it. It wasn't too heavy—all it held was the frames and about thirty thousand bees weighing about six pounds—so I could balance it until Dave came down. Then he took the hive and descended the other ladder and put the hive in the car trunk.

As we were about to leave the man said, "If we have any more bees, we'll remember you."

"And don't expect the fire chief to come," Dave quipped.

The man's laughter echoed in our ears as we left.

Bees in a Chimney

One Saturday afternoon the telephone rang. There was a frantic man on the line. His voice was choked with emotion as he said, "The fire department says you are beekeepers and can get bees out of my chimney. Is that so?"

"Yes, I am a beekeeper and I can try to get the bees."

He interrupted. "Trying to get the bees isn't enough, you'll have to get them."

"Bees in a chimney are a difficult situation—will you tell me more about it?"

"They're up there somewhere. I built a fire in the fireplace and tried to smoke them out, but the house got filled with smoke and I called the fire department. They came out with CO_2 but that didn't help because there was no fire in the house. Then they used fans and blew soot all over the place, but the bees are still up there and the house is a mess. Please, will you hurry and come over and do something about it?"

"Where do you live?" I asked.

He told me and I said we would be right over. Dave was in his room, so I went and told him about it.

"Wow," he said. "Sounds like we are going to have to kill them if they are entrenched in the chimney. I better go get some bee exterminator to take along."

"It sounds like a pretty messed-up situation and we can't get them, so I guess you're right."

We never kill bees except in emergencies, and this sounded like one. Most professional exterminators do not like to kill bees either unless there is no other solution, because they realize the ecological value of bees.

When we arrived at the house we found a middle-aged man and his elderly mother, both very upset. Dave went to the basement with the man and I stayed in the living room with his mother, who was extremely nervous.

"George has a lot of antiques down there, including books, and he is afraid they are all ruined. He is so upset I'm afraid he may have a stroke because the books, furniture, and statues are all covered with soot. The fire department certainly made a mess, although they tried hard enough to help." She wiped her furrowed brow with a small towel.

I was afraid she was the one who might have the stroke, she was so white and trembling. "Don't worry anymore, I'm sure Dave can get rid of the bees. Then you can vacuum up the soot and wash the things with mild soap and water."

Just then Dave and the man came upstairs and Dave said, "I put the newspapers down in the fireplace under the chimney and opened the windows. Now I'm going up on the roof and spray this down on the bees in the chimney. It may not spray far enough but it will fill the area and fall on them; I closed the damper, so there will be no draft to flow it back up."

Dave climbed a ladder and sprayed the beekiller down the chimney. In about half an hour he returned to the basement to see the results, and the man went with him.

When they came up Dave said, "You should have seen what happened when I opened the damper! All the dead bees fell out on the newspaper." He held up the bundle.

"I searched through them for the queen—here she is." He held his hand out and there in his palm was a beautiful dead queen. He put her in a small bottle the woman got and slipped it into his pocket.

The man was very grateful and gave Dave fifteen dollars for his work. Dave didn't want to take it because he felt bad about having to kill the bees, but the man insisted. "You must take it, you did a magnificent job. If only I had called you first. Besides, it is worth far more to me than money, and what you said about vacuuming and cleaning with soap and water will probably take care of the rest." He was wiping his eyes, which were quite red, and I could see that he had been crying.

"Thank you, and I do hope you get everything cleaned up OK."

Dave put the paper full of bees in the trash and on the way home he said, "I was never so sorry for anyone as I was for that man. He cried like a kid about all the soot all over his antiques. He must have thousands of dollars worth down there and he thought they were all ruined."

"And you told him how to clean them. I told his mother too and I do hope it works."

We drove silently for a while; then Dave said, "You know, I wonder why more chimneys don't get bees in them. There are so many of them and they are so handy, yet this is the first time we've been called about one."

"Maybe it is because most of them smell too smoky, wouldn't that keep bees away? If many of them had to be treated as this one did we'd have to be killing bees all the time, and I wouldn't like that, would you?"

"I'll never do it again," Dave said. "I still feel sick."

As it turned out we did have to kill another swarm, and very soon too. A neighbor of ours owns an old three-story home and rents apartments. Bees had gone in between the second-story porch and the wall of the building, which is brick. An elderly tenent feared they would get into her apartment and was threatening to move if something wasn't done. The owner came and asked our help.

Dave went over to look, then came back and said, "I'll have to kill them. Jack wants them out immediately so I don't have time to try anything else." He bought another spray can of bee killer, climbed a ladder, and sprayed it into the hole where bees were going and coming. That was the end of them.

These bee killings shook us both and we resolved never to accept any more such calls. These two were ones we could not refuse because people were seriously involved. We never again killed bees; although some people who called wanted us to kill bees under special circumstances, we either managed to take the bees or advised the people to call an exterminator.

Bee Trouble

In midsummer we went out to the apiary to inspect the supers. As we were getting out of the car, the man who owned the property came toward us waving a hand to get our attention. He lived in a large white house at the front and was elderly and very thin.

His words greatly surprised us, for he said, "I want you to move your bees away from here as soon as you can."

"Have we done something we shouldn't have?" Dave asked.

"No, no, nothing," he replied, shaking his head.

"Is Mr. Johns to take his away too?" I asked, thinking that the man no longer wanted bees on his place.

"No, his will stay, I just want you to move yours and as soon as you can." He turned and went into the house.

Dave and I looked at each other in disbelief. "So Johns keeps his bees here. Do you suppose he is in back of this?" he asked.

"I can't imagine why he would do a thing like that."

"I haven't seen him since the bee deal, but I thought we were friends. There is certainly enough forage for his bees and ours."

"Nine of those colonies are the ones we took for him," I remarked. Oddly, we never heard from him again, so it remained a puzzle.

Since our other moving my sister had purchased two-and-one-half acres farther out where she kept some horses, so I arranged to put the bees in a corner. She thought it would be better if they adjoined the corral, so we went to work preparing it by building a rail fence enclosure to keep the horses from going among the hives and being stung.

Dave dug eight postholes and set the rough cedar posts and we nailed the rails to them. This took one weekend and I became a great admirer of Lincoln even though we didn't have to split the rails as he did. We had to build only two sides, the corral forming the third side and the acreage fence the fourth. We planted a tree in the middle of our twelve-by-twenty-foot area so there would be shade during hot weather.

The next weekend we went to the bees on Friday night, stuffed the hive entrances, and moved two colonies to our new apiary. It was too dark to do more, so early on Saturday morning we began to move the others and were through by noon. Our car looked odd on the highway, two tall white boxes in the trunk and the top standing up. When the hives were all set our apiary looked very nice with two rows of hives, staggered for ease in handling. We set the hives on small stands so they would be off the ground; the hives were three feet apart so bees would not enter the wrong hive.

The next year my sister said, "The bees bother me when I am working out here and Alma, who rides with me, is afraid of them. She says she will quit riding because the bees are so close. Could you move them down to a far corner?"

"I guess so. The material we have will be enough, since there will be two sides of the acreage fence. It will take a couple of weekends to do it, just as it did to bring them here. Will that be OK?" She said that it would, and thanked me.

Dave and I moved the fence and then the bees a long block away and downhill to a corner of the acreage, but our tree still stood up by the corral.

"What do you say we move the tree?" Dave asked early the next spring.

To me it seemed a tremendous task, but we drove out to look at the situation. Dave took a shovel and dug a big hole in the middle of the apiary to accommodate the tree, as I watched.

"It's easy digging, real sandy," he said. "I don't think it will be too difficult to move the tree and this place will be ready for it."

Then we went up to the tree and Dave began to dig in a circle around it. It wasn't too difficult except that we had to make a ball of the roots and then dig under to get the taproot loose.

"There it is," Dave said, felling backwards while the taproot came free as he pulled on it. We managed to lift the tree to a wheelbarrow and take it down to the apiary.

Amazingly, it grew and by fall was a nice tree. We had a dozen colonies then, all but two in hives that I had made and topped with supers from the same carpenter. They all faced south away from prevailing winds and for sun in winter.

Once a week we took water out for the bees, since there was no stream nearby. I bought a three-gallon chicken waterer, constructed on the "inverted glass on a saucer" principle. It is galvenized metal and has a valve at the bottom that lets water out into the rim as it is used. The moat around the bottom must be filled with strips of old sheeting to keep bees from drowning in the water. They congregate all over the cloths and suck water from them or drink from the tiny pools formed in the folds.

Requeening Colonies

After refilling the waterer one day I saw a queen struggling in the grass. She was evidently injured, since she was down in the grass instead of flying.

I called to Dave, "Here's a queen on the ground. I wonder if she is from one of our hives."

Dave picked her up in his hand and we saw that her abdomen was slightly caved in. "I'll go see if I can find out," he said and he listened for a time at each hive with his ear close to a side. At the first one he said, "This one is humming happily," and went on to the next one. At number six a dismayed look came over his face and he said, "This is the one. They are wailing in a minor tone so their queen must be gone."

I went over and listened. "You are so right. They certainly are wailing."

"What do we do?" Dave asked.

"We can do one of several things. Send for a queen by mail, double up with another colony, or look for a queen cell in another hive."

The first suggestion appealed to him. "It is really too big to incorporate with another colony, don't you think? And it's a bit early for queen cells. Why don't we send for a queen?"

"It would be an experience and something we haven't done," I said. So when we got home Dave sent for an Italian queen.

She arrived in a few days in a little box, ½x1x3 inches, a bit larger than a penny matchbox. The top had three slightly overlapping holes covered with screen wire and we could see the queen and three attendants inside. There was a hole about the size of a pencil at each end with stoppers of hard candy or crystallized honey in them. This was not for her to eat: a queen doesn't feed herself but is fed by special bees in the colony. These bits of candy were to be eaten by bees in our colony, who would become used to her as they ate and freed her.

Dave hung the little case in the hive between two of the center frames and put the cover on. The next day we went out to check, as the book said that in twenty-four hours the bees would have released her. When we opened the hive we saw that instead of releasing her, the bees had stung her to death.

"Could the book be wrong or did we make a mistake?" Dave wondered.

"Maybe you should examine the frames and see if there is a queen cell anywhere," I said.

"There surely are queen cells in one, three cells in fact. They must have made them while we were waiting for the queen to arrive. So that was why they stung her to death, they wanted to make a queen for themselves and were doing it."

"There must have been eggs or larvae less than three days old when we saw the queen in the grass. It is a very active colony," I said.

"They aren't wailing anymore. I wonder if they had already stopped when I put the queen in. I was so interested in what I was doing that I forgot to listen."

"Oh, well, the colony is safe now and our worries are over."

Later on we discovered another colony without a queen and wailing. This was a newly taken swarm that had a queen when we took it—we had seen her.

"She must have been lost on her mating trip. There is no brood in the hive, so we have a problem," I said.

"We could give them a frame from one of our strong colonies," Dave suggested. "Then we will be sure it has eggs and larvae."

He looked through some of the hives until he found what he thought was the right sort of frame, shook all the bees off and traded it for an empty one from the new colony. It had all the necessities for queen making: eggs, larvae, sealed brood, honey and pollen, but no queen cell.

"Some larvae must be less than three days old, so they can go right to work," Dave said. "And even if there aren't, they will have eggs and can go from there, but it will take three days longer."

"She will be a fine queen," I said as I saw the well-filled comb with a large oval of brood. "And the new swarm will not only have a queen, they will have all the sealed brood too."

We checked the colony in a few days and it was humming happily. Dave opened the hive, very carefully removed the frame we had given them, and said excitedly, "Right on! There are three queen cells on it, so they will be OK."

Every bee in a colony seems to know whether the queen is missing or safe in the hive, and when a new queen is in the making. The first queen to hatch will be the one for the hive; she will sting the others to death, or the bees will tear the other cells open and destroy the nymph.

Another time we found a queenless colony early in the spring. It was a small colony that we had taken late in June. The bees had wintered well but were wailing. Finding a queenless colony in early spring is really a problem, since it is too far from swarming time for any of the colonies to be making queen cells. We could have given them a frame of brood as we had done for some before, but Dave wanted to vary our experiences.

When I told him about it he came over and listened. "Sure enough. That's the small swarm we thought might not live through the winter. What do you say we combine it with another colony?"

"That's a good idea, since there are no queen cells to be had, and we've never tried combining colonies."

"I wouldn't want to try ordering a queen again, would you?" he asked.

"No, I don't think so. But which colony should we use for combining?"

We stood thinking. Then I said, "What about the one that wintered in the backyard? It isn't a very big colony either."

Dave read the bee book for instructions, "It says here, 'Combine colonies by putting one on top of the other, separated only by a sheet of newspaper. They will become acquainted, eat through the paper, and become one colony.' "

That evening we went out and brought the queenless colony home. Dave said, "I will combine them while you are fixing supper."

But the book left something out or Dave didn't read all of it. After a day or so I heard an angry buzzing in the upper hive and called Dave to come and listen.

"Wow! They sound mad and are either fighting or want out. I'll bet they are thirsty, penned up all this while. What should we do?"

"You lift the upper hive ever so little and I'll pull the paper out and hope for the best," I said.

Then Dave looked at the paper. "Not a bite in it! I don't dig it."

"I dig it," I said. "You should have punched some tiny holes with a toothpick in the paper before you put it over the lower hive, with the little rough places up so the colony to be introduced could have had something to pull on to get out."

"I'll bet that would have done it—why didn't you tell me?"

"You never asked me. Besides, I never thought of it myself until I saw the paper as I pulled it out just now. Oh well, it is done now, next time we will know about it."

Dave listened by the hives, "The angry sound is gone. Far out! They must have become acquainted even through the paper, because the upper colony seems to be accepted."

"Remember, they have a hive of honey and no queen and that should help a lot." Later we took the combined colonies to the apiary.

Great caution is needed in giving a strange queen to a colony, whether she is an impregnated queen purchased from a bee raiser or is secured from one's own apiary. If she is introduced too soon, within twelve hours of the removal or loss of the reigning queen, the bees may surround her, seizing her and keeping her captive in an impenatrable cluster.

This is called balling. The ball may be as large as a fist and so compact that it cannot be easily scattered. She may be rescued by dropping the ball into a basin of water, then carefully picking her out.

If eighteen hours are allowed to elapse before introducing a stranger queen, she may be treated in the same way, but the bees leave her sooner and the cluster is not so tight. They gradually disperse and she is liberated at last to move languidly about. She may die in a few minutes, or may survive this treatment and live to reign in the hive.

Reigning does not mean that the queen really rules the colony. She has little to do with the business of the colony other than to see that mostly worker cells are built in the combs, then to deposit eggs in them. She is the hive mother, adored and fondled by all the workers—but not by the drones, who ignore her in the hive.

The way a stranger queen is treated by a colony, when they have been queenless, depends upon the honey flow. In a good season queens have been introduced to colonies and immediately accepted because the bees were extremely busy and knew that there was a great need for brood. But I would not recommend introducing a valuable queen in this way.

To introduce a strange queen to a colony successfully, certain conditions must be met. The colony must be absolutely queenless for twenty-four hours with no facilities to make a new queen. Bees recognize one another by scent; the scent of a strange queen alarms them. A laying queen has a peculiar odor that pervades the hive. Her presence is known by all the bees. When a new queen is introduced, if the beekeeper can arrange for the bees to become acquainted with her before she is released, she may be accepted.

Correct introduction is accomplished by placing the queen in a small cage made of wood and wire cloth, which is hung between the combs in the most populated part of the hive near the brood and allowed to remain there for twenty-four to forty-eight hours. The wires should not be closer than twelve to the inch, permitting the bees to feed her readily through them. They will become acquainted with the imprisoned queen by thrusting their antennae through the openings and will quiet down as though she had her liberty. Now she must be released or else the cage must have a candy or granulated honey plug which they can eat to release her.

After she is released the hive should be left alone for two or three days before it is checked to see if she is present and laying. Any unusual disturbance during this time, such as light in the hive, may bother the bees to the extent that they will kill the queen.

The queen to be introduced to a colony may be an impregnated one purchased from a bee raiser or a virgin queen from one's own apiary. The reason for introducing a queen in a cage hung between the combs is that there is no room between the combs for the bees to ball her when she is released. Colonies have been know to ball their own queen if dissatisfied with her or if she is not productive. They will crowd around her suddenly, pull off her wings and legs, and sting or suffocate her. Then they will replace her.

The foregoing applies to introducing an impregnated queen. Bees that have had a fertile queen are quite reluctant to accept a virgin. It takes an expert to introduce a virgin queen and secure a good reception for her. The best time to introduce her is soon after her birth when she can crawl readily, and during the honey flow. If it is done during swarming season the bees may swarm with the queen given them and leave the hive still queenless. The apiarist should watch the colony closely for a few days. A virgin queen looks quite different from an impregnated one: she is slender with a small abdomen and quick motions. She runs about and will almost fly over the combs when trying to hide from light. This disturbs the bees and may cause them to attack her, sting her, and drag her out of the hive as they would an intruding worker.

One recommendation is to daub her with honey and introduce her without touching her with your fingers. A safe way is to introduce her to a small nucleus of young bees who have been queenless for ten hours. A virgin queen is fuzzy, like a young bee, which is another reason that a colony may not accept her.

A virgin may also be introduced by putting two or three frames, with bees from a queenless colony, into an empty hive

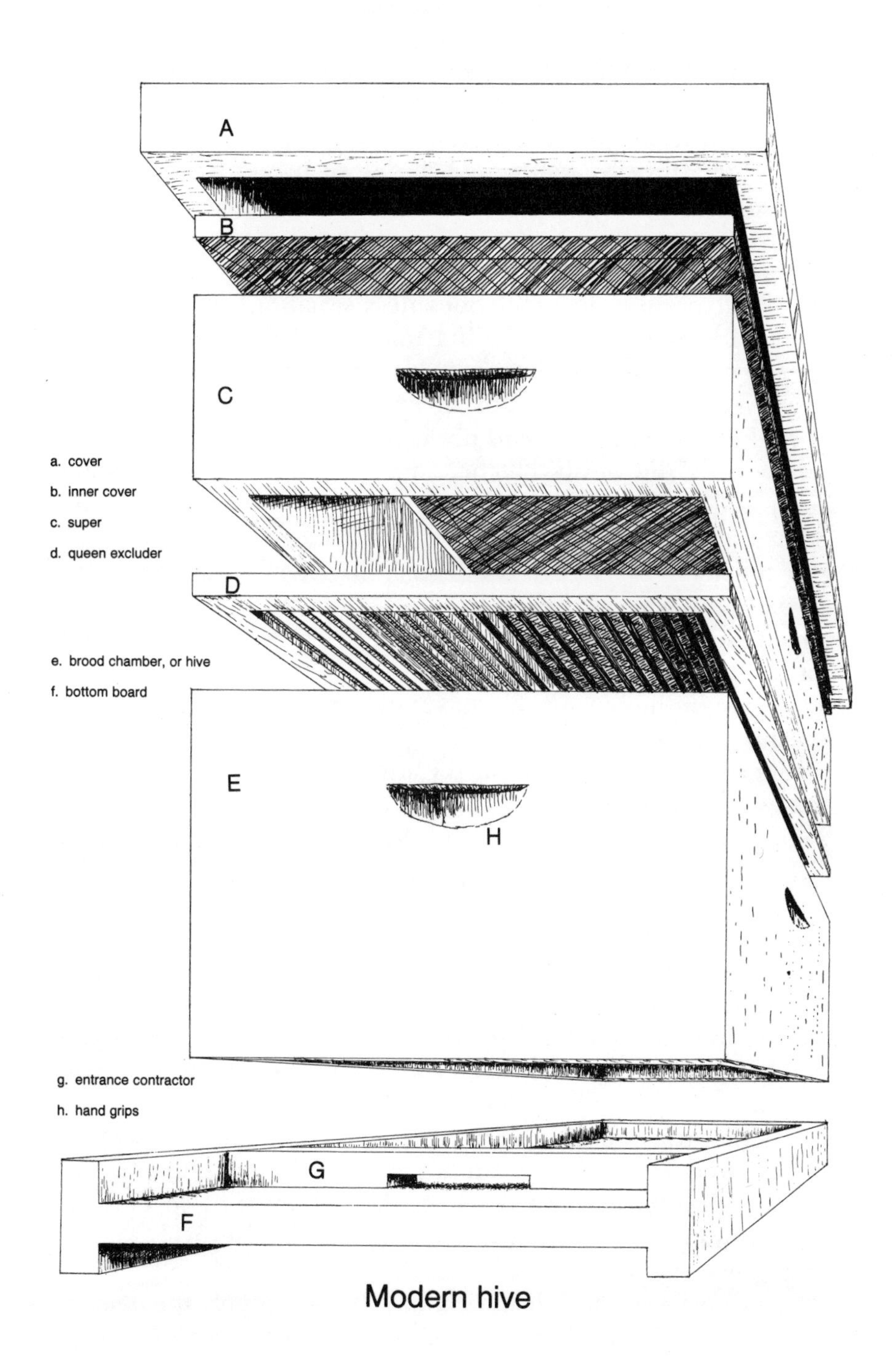
A
B
C
a. cover
b. inner cover
c. super
d. queen excluder
D
e. brood chamber, or hive
f. bottom board
E
H
g. entrance contractor
h. hand grips
G
F
Modern hive

along with her. When the bees have accepted her, return the frames to the colony and she will be accepted.

This all seems so chancy that I prefer to secure a virgin queen for a hive by removing a frame of brood from another hive and giving it to them to create their own queen. This will take from ten to fourteen days, depending on what the bees do; then there is no question of them refusing her.

I usually want to have the queenless situation solved quickly. If I could locate a frame with a queen cell in another hive, I gave that to them and the queen would hatch in about a week. Bees will always take combs of brood and whatever hatches from them when they would not take an adult queen.

However, the newly hatched queen must take her mating flight and if there are bee predators in the area, there is the chance of losing her.

A third way to secure a queen is to check other hives for intent to swarm, as shown by the presence of several sealed queen cells. All that is necessary is to use a queen trap and watch the hive. If the apiary is far from home, as mine is, I put the frame in the hive of the queenless colony at the time I find it. This will, of course, give a virgin queen. The queen and a number of drones will be in the queen trap when the swarm comes out and the queen may be given to the queenless colony. She will, in all probability, be an older impregnated queen and will not need to go on a mating flight. She may be more readily accepted in a new hive.

This is mostly for a new colony with no brood, although we have used the placing of a frame with a queen cell in an established colony if it was queenless with no queen cell of its own. If an established colony loses their queen they usually can create another. They may even supersede her if for any reason they are not satisfied with her.

Keeping track of the queens is a big task for the beekeeper. We would open a hive to examine the brood frames, in order to replace an unproductive queen if the colony showed definite signs of waning. Also we would check whether they were superseding her.

Some apiarists use chloroform or ether to stupefy the bees of a mutinous colony that persists in refusing to accept a stranger queen and show it by angrily surrounding the cage in which she is imprisoned. When a queen is removed the bees will run in and out for some time in search of her. They were aware of her scent in the hive and the absence of it causes them to hunt for her. Her scent is so attractive to them that the slightest touch of her anywhere in the hive will attract them. They will even run inquiringly over our hands after we have caught her.

In hunting for a queen in the hive it must be remembered that she is to be found on the combs of brood unless she has been frightened away. An Italian queen may be found in five minutes after opening the hive because she is large and golden and not easily frightened. A queen of common or hybrid bees is more difficult to find, since her bees rush about as soon as the hive is opened. If she cannot be found it may be necessary to shake all frames over a sheet in front of the hive and pick her up as she falls. Remember not to hold her tightly when she is picked up, as she is easily injured. This poses no danger to the beekeeper, by the way: he could crush the queen to death and she still would not sting.

Bees on a Car Grille

One Saturday Dave answered the phone and I knew from his conversation it was a call to pick up bees. When he hung up he whistled and said, "Guess what now!"

"I've no idea, darling."

"There are bees on a car again, but not inside. It is out on Parker Road by a dairy. The woman said when she came out to get in her car to go home she saw the bees—she works at the dairy."

"It is lunchtime, can we go later?"

"I'm afraid not, she was very excited and wants us to come right away."

"All right," I said, "you put the things in the car and I'll make us each a sandwich and we can go."

I made two husky sandwiches and poured two glasses of milk. I ate my sandwich while he loaded the car, he drank his milk when he came in and ate his sandwich while I drove to the dairy. Dave was not sixteen yet, so I did the driving, but we were partners; so he did most of the bee work while I assisted.

When we arrived at the dairy he went in to find the woman and she came out and took us to the car. There were bees on the grille and the headlights.

"Far out!" Dave said and went to get a hive. The woman was very nervous and asked, "How long will it take? My husband will soon be home wanting lunch."

"I really don't know, so why don't you wait inside and we will come and tell you when we have them. It is hot out here and we usually discourage people from watching because they might be stung."

Dave put the hive under the bumper and began to brush bees onto a frame which he put into the hive. But the bees came right back to the car front and we sat watching for the queen while they milled around over the grille and headlights.

"I hope the queen hasn't gone inside somewhere like she did on that other car," Dave said. "I'll never forget that, will you?"

"Indeed I won't, but that time the bees were inside the car. Now they are outside and on the front."

Just then the queen ran across the bees on the headlight. We both saw her. Dave reached for her but missed, so we waited and watched again. So she was not under the bees as usual and we hoped she would come by again. In a little while she ran by and Dave was prepared this time and picked her off. I brushed some bees onto a frame and put it into the hive and Dave dropped the queen on it. Soon we had all the bees in the hive and they stayed this time.

Dave breathed a sigh of relief. "That was easy, but I didn't think it was going to be. If I had failed to get her she might have disappeared somewhere and we wouldn't have these bees."

While Dave made the hive ready to take home, I went in and told the woman we had the bees. She was as relieved as Dave.

"What do I owe you?" she asked.

"Not a thing. We keep the bees and are glad to be of service."

There was no question of waiting for scouts this time. If there were any there would be no car there when they came back and they would return to the colony from which they had come.

We were interested in people's reaction to bees; they were mostly afraid but always extremely curious.

Managing Bees

Since in modern management we take many liberties with bees it is important to know how to perform all operations without arousing their anger. Many people are greatly astonished when they see an apiarist open hive after hive with the use of smoke, remove combs covered with bees and shake them back into or in front of a hive, then form new colonies, find a queen or queen cell or transfer the bees with all their stores to another hive—in short, to do all the necessary handling as though bees were harmless flies. Some people have even asked if the colonies we were working on had been trained to accept such work.

It must be evident to our thinking that the Creator intended bees, as well as our domesticated horses and cows, for the comfort of man. Honey was the only sweet that early man had and it was gained at great hazards. Nowadays, with modern methods and careful handling, it is easy to obtain. Since the honeybee can store far more honey than it uses, man has learned, with certain precautions, to obtain a share without harming the colony.

Learning that a heavily loaded bee never volunteers an attack but acts solely on the defensive must have been a great help. The first man who attempted to hive a swarm was no doubt agreeably surprised at the ease with which it was accomplished.

He did not know that the bees had gorged on honey before they left their home, and so were mild. With this honey, bees start to build comb in their new home; they do not become hungry in flight or if several days of stormy weather should follow their migration.

A heavy load of honey so distends the bee's abdomen that it is impossible to bend at the correct angle to sting. Nothing short of abuse can get them to part with their supply, so the hiving of a swarm would be riskless if it were not for some improvident bees who decide, too late to gorge, to accompany the swarm.

These bees are so unhappy and filled with hate for anyone daring to meddle with the swarm that they will sting even though they die for it. If the whole swarm were as ferocious, no one would dare to hive them without a beeproof coat of mail and we would still be getting our honey from hollow trees and clefts in rocks. Bees would never have been domesticated.

So much for bees in a swarm; now for those at home in their hive. The best tranquilizer for bees is smoke, so a smoker is indispensable to an apiarist in handling an established colony; it renders even the most fiery colony as docile as a swarm. Bees are frightened and retreat before the smoke, seeming to imagine that their honey will be taken away, so they gorge to their capacity.

The smoker may be filled with small pieces of dry wood, dead leaves, wood shavings, or rags. We found old pieces of outing flannel to be very good. Some apiarists use "saltpeter rags" in a smoker: the least spark will light them and they continue to burn and ignite pieces of wood.

To make saltpeter rags, a quarter pound of saltpeter is dissolved in a gallon of water and the rags are dipped into this solution, wrung out, and dried; then they are ready for use in the smoker, and may be stored for use. Any pieces of wood added to them will at once be ignited into a good fire.

Personally I have never cared for this method because it creates more heat and fire than is wanted, but less smoke. Fire belching from the spout of a smoker can scorch bees, so we were careful never to have a flame in the smoker after it was well started, but only smoldering contents that put out lots of smoke to quiet the bees. A continued tapping on the hive is said to break up the protective program of the hive and the bees will not sting. It is claimed that the first blow angers the bees but continous blows seem to quiet them. We always found that anything knocking against the hive brought out angry guards.

Smoke from a smoker is good, but improvised smoke may be dangerous. Tobacco smoke is harmful to bees and unconfined smoke or fire can cause dire results. One man endeavoring to secure a swarm lodged in the wall of a frame house made a wad of paper, placed it against the wall, added oily rags, and applied a match. The fire department came promptly, but the damage ran into the thousands and the bees had departed.

The most recent development is an aerosol smoke bomb that puffs nicely when a button is pressed. Also, it has been discovered that bees stop motionless in the hive as long as they are exposed to high intensity sound waves of from 600 to 800 cycles per second—about one-and-one-half octaves above middle C. Such waves from a loudspeaker placed about one yard from the hive can halt its entire activity. This is too new a discovery to have been used widely yet.

The bellows smoker is excellent and was used long ago in Europe, but the early version was not practical since it could not be operated with one hand. Today's smoker has a bellows on the side of the firebox and is easily operated with one hand, leaving the other free for work in the hive. By setting the smoker upright when not in use a draft is maintained and a good supply of smoke can be had all day with refueling from time to time.

In addition to the problem of calming the hive and field bees, there are the guards to consider. They remain near the entrance and go out on the defensive at the least disturbance around the hive, such as prying off the cover to use the smoker. So the careful apiarist puts a few puffs of smoke in the entrance before he removes the cover.

Mild Italian bees need only a puff or two of smoke, but Cyprians cannot be controlled without a cloud of it. Bees can be handled at all times but are quietest in the middle of the day, when the older or field bees, which are the crossest, are out in the fields gathering nectar—except on a cloudy day, when they are in the hive and are most irritable. Field bees returning loaded with nectar are not cross, but those leaving are. During a heavy honey flow nearly all bees are full of nectar and colonies can then be handled with little or no smoke.

Gloves are worn to protect the hands, and a veil for the face (the most vulnerable place). Some apiarists rub liquids on face and hands for protection; a compound of oil of cloves, wintergreen, turpentine, and other volatile oils, plus ether or chloroform, is said to be effective. However, some apiarists complain that their bees did not mind it and stung as usual, while the beekeeper got blisters on hands and faces after use.

A carbolic acid solution can be bottled and kept for use. Mix one-and-one-half ounces of carbolic acid with an equal amount of glycerin, add a pint of water, and shake well. Use as a lotion to rub on hands and face, or dip a cloth in it, wring out, and place over the hive as soon as the cover is removed, to quiet the bees. The liquid may also be sprayed among the bees with an atomizer; it evaporates and leaves no odor.

Still another method, especially for clearing bees from a super, is to cover it with a carbolic acid cloth: they will be out in fifteen to twenty minutes. We took a bottle of pure carbolic acid crystals, placing it in hot water until the crystals were melted, then diluted with an equal amount of water. This is corrosive to fingers and skin, so wash at once after using. To use on supers, take a screen on a frame the size of the super, tack several layers of cheesecloth on top, and cover with oilcloth. Put the carbolic acid solution in an atomizer or spray bottle and spray the cheesecloth just wet, not dripping, and set it over the super. We used this only once, as we thought it might taint the honey.

There was always a question in our minds about the effectiveness of these chemicals, so having gloves, veils, and a smoker, we kept it simple and used them. They were sure protection with no fuss or guesswork, and we felt safe. If heavy bee gloves seem clumsy, durable rubber gloves may be used: they are impenatrable to bee-stings, yet light enough not to interfere with the work to be done.

Stings on the hands cause little pain or swelling, while stings on the face are quite painful and a grotesque appearance is caused by the swelling. Thus I wear a veil.

One type of veil is meat safe, made of four panels of screen wire bound with heavy tape, joined together and stitched to a crownpiece above and a neckcloth below. The crown has a drawstring to tie snugly around the crown of a hat and the neckpiece ties around over the shirt collar. Another type consists of wire encircling the head and face with net above and below; the veil slips around the crown of a hat and ties around the neck.

My veil is simply an army surplus mosquito hat of dark green net with a taffeta crown and a drawstring to tie around the neck. Veils are frowned on by some beekeepers, but in a large apiary they are a necessity, as nearby colonies may be aroused accidentally. Veils should always be black or dark green: other colors are difficult to see clearly through.

Woolen clothes are more objectionable to bees than cotton or linen, because wool comes from animals, which bees dislike, while cotton and linen resemble the plants with which bees are familiar. If bees alight on beard or hair they will sting if they can reach the skin.

In opening a hive, there is little danger from the bees on the combs that are exposed to the light. It is not merely the sudden admission of light, but its coming from an unexpected direction, that calms the bees. The light followed by a quick puff of smoke sets them to eating honey and renders them easy to handle. Too much smoke may drive them from the hive and cause them to get in the way of the apiarist. The bees near the top of the hive are already gorged with honey in order to make wax, and are no threat.

Let all motions about the hive be slow and gentle. It is important not to crush or kill even one bee in opening a hive, for bees communicate their sensations with magic speed. A colony can be aroused to fury by an angry note of a single bee or by her death wail. It also irritates bees to breathe on them or to jar the combs.

Removing frames from the hive or super requires great care: they are glued down to the rabbets with propolis and must be pried up gently. Push the third frame a little nearer to the fourth frame after gently loosening it at both ends. Then push the second frame as near the third frame as possible in order to make room to remove the first one without breaking the comb or injuring any bees. If the beekeeper has a frame grip, he should carefully place it at the middle of the top of the frame and gently lift it out, otherwise he must lift the frame out by hand.

As combs are made slightly waving it is impossible to remove a frame successfully without first making room for it. Combs built in frames are straight, but bees still sometimes connect them. Occasionally we found two combs so built together in a super that it was not possible to remove one at a time. We removed them as a unit and separated them later.

If one frame is desired, room must be made on each side by moving the adjoining frames as much as possible. If all combs are to be checked, set the first frame carefully on end near the hive; then the second frame can be moved into the vacant place and examined. Proceed in the same way until all have been examined. Use the same procedure when removing frames from a super, so as not to break the combs. To protect them from wind and robbers, frames may also be put into an empty hive as they are removed; be sure to keep them in order.

In returning frames to the brood chamber care must be taken not to crush bees between them and the rabbets on which they rest. Frames must be put in slowly so that a bee, on feeling a slight pressure, has time to creep out from under and not be hurt.

The frames must be returned in the same order and position as they were taken out, with the brood at the front of the hive and honey at the back; for bees always breed in front of their stores in order to better defend them against robbers.

When closing a hive, the inner cover should be slid slowly over the frames and the outer cover replaced. If bees are unskillfully handled they may "compass about" the worker with savage ferocity, and woe to him if they find an unprotected spot. The unskilled beekeeper kills bees and that promotes anger. We never strike at bees, for if one is hit she buzzes in a peculiar way and others come to avenge her. If this continues, hundreds of bees will attack. The best thing to do then is to retreat to a building, hide in the bushes, or lie in the grass face down until the bees leave. With our veils, smoker, and careful ways, we retreat very little and keep calm.

It goes without saying that a beekeeper must not offend his bees; he takes care to be clean, sweet, sober, and quiet—and familiar, so they will know him from all others and he can work among them successfully. To stand still before a hive is to invite stings and to blow on bees is to cause them to try to sting the source.

Some authorities advocate examining hives at regular intervals, others say the less the bees are disturbed or disorganized the better they work and the larger the honey crop. We agree with the latter; we open our hives only when the bees' welfare is in question. We have not done any superseding of queens except to replace a German queen with an Italian one, in order to obtain a more docile, easily handled colony in the course of natural replacement. German bees are easily angered and it was difficult for us to handle them. We found that the bees pretty well took care of superseding old or unproductive queens; however, if we found a colony queenless we supplied a queen.

We harvested forty pounds of honey per colony; on one occasion, eighty pounds from a colony where we had prevented swarming by removing queen cells and giving a second brood chamber. It was a double brood colony with one queen and filled two supers with honey. As cold weather came we found that the bees had vacated the upper chamber when the laying season was over, so Dave removed it. Later we used it to put a swarm in, for good straight worker comb from an abandoned hive or a dead colony can be used to good advantage for a new swarm. It will free them from the need to build combs and brood can be raised at once.

Bees have an acute sense of smell; unpleasant odors provoke them. Even though we prize perfumes, bees find them disagreeable and may sting the wearer. The smell of their own venom irritates them too; where one bee has stung, others will attempt to. So if one's clothing is stung it is best to depart or to puff some smoke on it to conceal the odor. Strong body odor is equally objectionable, and bees have a special dislike for those who are not neat.

A horse wandering among hives is often killed; for instead of running if stung, it will plunge and kick until it overturns a hive. Or it may roll, with the same result, until tens of thousands of bees sting it to death. In one such case, although the carcass remained unburied for several days, neither dogs, crows, buzzards, nor any other scavengers attempted to feed on it, so great was the amount of poison put into it by the bees.

In the spring we inspect the hives for strength of colony and supply of food; since no flowers are in bloom, yet the bees will need feeding if their food supply is low. However, we have never had to feed an established colony that had wintered well, since we never took honey from the brood chamber. Our apiary is located on rather sandy soil, and I check the hives for slope in spring and fall. A hive should have a slope of about one inch from back to front so that rain in summer and melting snow in winter will run off the alighting board in front and not back in the hive under the bees. I check with a level: if a hive has not enought slope, Dave raises it at the back and I slip a piece of wood of the right width under to give the correct slope.

Bees should be protected from prevailing winds as they approach their hives. Dave and I wove some thin plywood panels into the wire fence on the north side. Our apiary is in a corner of the acreage and the north and east side fences are wire. The south and west sides are the rail fences that Dave and I put up.

In May and June we took swarms that came from our colonies or took steps to prevent swarming. Since our apiary was over five miles from home, swarming occurred anyway. Sometimes the swarm settled on a fence post or the limb of our tree and we hived it; at other times it just went away. We set the supers in June and harvested the honey in September, when we checked the colonies for strength. There is some nectar brought in until frost, but we allowed the bees to store it for winter in the brood chamber. One or two side combs should be stocked with honey and the center ones should still contain eggs, larvae, and the sealed brood with pollen and honey at the back.

If bees need to be fed, use two parts of sugar to one part of boiling water in a five-pound tin with the lid on tight. Twenty or thirty nail punctures are made in the lid and the can inverted on a queen excluder over the frames with an empty super for cover on top. All feeding should be over by mid-October.

Bees winter on their permanent location in this area without any banking or packing. We remove any supers we may have left with honey for the bees and close the entrance to about one-fourth of its summer size with an entrance contractor or block, which conserves the colonies' heat and prevents the entrance of mice. All work is done on warm days between ten in the morning and three in the afternoon.

We still wear gloves and veils and use the smoker; for bees can become very angry when their brood chamber is opened. If Dave must look for a queen he works very carefully so as not to knock hundreds of bees head over heels and create extra excitement. He usually finds her on one of the inner frames among the brood.

Bees do not hibernate in winter: they are dormant, clumped on the combs, eating only a small amount of honey from December to March. No one knows the exact amount of honey needed to winter a colony; it depends on the size of the colony and the severity of the weather. One apiarist has estimated it at twenty-five pounds; but as we left all the honey in the brood chamber, we were not concerned. Dave would heft them and estimated they contained about forty pounds each.

The natural heat of trees protects bees in the wild during cold weather; bees in a hive keep warm by clustering. The bees in a cluster overlap like shingles, each bee with her head under the abdomen of the one above her; those on the honey level eat. If the cold is intense there is a constant trembling motion like a person shivering, to create more heat; the fanning of wings sends warmth to the bottom of the frame to warm those lowest on the cluster. When the temperature is 57° or less in the cluster, the outer temperature is around 43° and the bees most exposed must rotate to the inner level for food and warmth.

On warmer days the bees move around more and the humming can be heard outside the hive. Bees have a greater adaptability to climate and environment than many other insects because of their clustering and ability to regulate the temperature inside the hive. Bees fly out at outside temperatures of 50° on sunny days if the wind is mild. If the hives are banked greatly to protect against the cold, there is great risk of dampness, which can be fatal.

Bees are sold in packages, nuclei, or colonies. The package is small, weighing two-and-one-half pounds and containing some ten thousand bees. The packages may or may not include a queen; if she is present she will be in a small container in the middle of the package, to prevent her from being crushed, and she will be an impregnated queen. Normally bees in the hive weigh five thousand to the pound but bees to be shipped are heavy with honey they have eaten to carry them through shipment. Drones are heavier, with only two thousand to the pound.

Bees bought as a nucleus will be on several combs with an impregnated queen and will need feeding immediately until established.

The best way to buy bees is as a colony from a local beekeeper, then they are already established as well as being acclimated.

We have a calendar which we follow in our beekeeping.

January and February. Bees are in repose; do nothing.

March. Bees are active but weather too cold to open hives. This is the time the queen starts to lay in the center of the cluster if hive temperature can be maintained at 90° to 95°. The colony may not increase in size: the number of young bees may be less than the number of bees that have died during the winter. The condition is critical until the end of March when the crisis is passed and the population may be about 10,000.

April. Bees are active and flying out; but until flowers appear, all food for the brood must come from winter stores. Consumption rises from a few ounces a week to several pounds. The colony with large stores is off to a good start. The bees clean the hive by carrying, pushing, or dragging out old wax cappings and dead bees. When spring flowers appear, growth proceeds at a tremendous rate. Now drones appear; check hive for brood and strength of colony.

May. Swarming time begins; prevent or take swarms.

June. Swarming continues. Nectar plentiful; set supers.

July and August. Swarming is usually over, nectar flow is great; check supers every two weeks.

September. Harvest honey and remove supers. Most flowers are gone. Cooler weather, field bees are mostly home. Queen lays sparsely.

October. Frost comes and bees cluster when temperature falls below 57°. Contract entrances for winter.

November and December. Bees are dormant.

Water should be available in all seasons except when it is very cold or snow is falling, as it is necessary for bees to dissolve honey that has crystallized in the combs, to digest pollen, and to prepare food for the larvae. Water is supplied daily to larvae; should bad weather prevent bees from bringing water in for a few days, the larvae may perish.

Some beekeepers pour water into empty comb cells or put it in a feeder frame not then needed for feeding. In winter, when there is little brood, the moisture that condenses on the walls of the hive is sufficient. If water is kept available outside, the bees will take advantage of a warm winter day to take it in and use what is needed. This available water keeps the colony happy and quiet and saves many a flight in search of it. If there is no water in early spring when the weather is still chilly with brisk winds, many bees will perish in search of it. As many as four-fifths of a pound of bees, or 4,000 bees, were missing from a weighed hive during a sudden spring storm.

From the first of April to the last of July our bees consumed more than fifty gallons of water: the greatest amount was two gallons one extremely hot day. However, bees being shipped survive best without water. I keep a five-gallon poultry waterer, the moat laced with strips of old sheets, constantly full and setting among the hives. It must be refilled every ten days to two weeks in the summer; however, field mice may drink from it, as fields surround our apiary. Another thing bees seem to need is salt; they often alight on our hands to lick the salty perspiration. They seem to need it especially during the early part of the breeding season and will throng a pan of salt from early morning until late in the evening. During this time the amount they consume is considerable but later they are indifferent to salt.

To conclude this chapter on bee management I add my most recent findings on bee-stings. Because of the constant threat of being stung even though we both had immunities, I called a doctor and asked for information. He said there is a vaccine for people who go into shock if stung. It is made from bee venom which is diluted and administered in a series of shots given over two-, three-, or four-month periods, given each year like hay-fever shots.

Swarming

In early spring, as soon as the combs in a hive can accommodate no more brood, bees prepare to swarm with their queen. They build queen cells about the time that drones appear outside the hive; then when the drones leave, the parent colony will have a new queen.

Drone cells filled with eggs are the first sign of swarming. These eggs will take twenty-four days to hatch and develop into drones; then there will be another fourteen days to mating maturity. Bees need drones at swarming time for the mating flight of the queen—and also if a queen needs to be superseded or is lost. Queen cells on the comb are the sign that swarming will begin in a few days.

The swarming of bees is one of the most beautiful sights of spring; most apiarists delight in it. Swarming usually starts in May and continues through June and July; swarms have even come out as late as September, but only because the hive is too small or lacks supplies. Instead of facing slow starvation the bees sally forth to quick destruction, with no home or stores and frost approaching. In general, the time of swarming depends upon the climate, the earliness of the season, and the strength of colonies.

Swarms hived in May are the best, as they have the summer

honey flow ahead and time to build comb and fill it with honey and brood. These are the swarms that bring a harvest for their keeper in the fall, along with established colonies. June swarms have less time, and later swarms may perish unless fed. Early swarms take the old queens, secondary ones will have virgins.

There seems to be no infallible way to predict when a swarm will leave, as bees change their minds if the weather is bad or the honey flow stops. The best indication is when only a few bees leave a strong colony on a clear day when other colonies are busily at work. If a swarm is to leave, work is almost stopped until later. If the weather is very sultry, a swarm may issue as early as 7:00 a.m., but the usual time is between 10:00 a.m. and 2:00 p.m. when the sun is high. A swarm occasionally comes out as late as 5:00 p.m., but an old queen is seldom guilty of such an indiscretion.

On the day of leaving, the queen's voice can be heard in an impatient zeep-zeep. She is very restless and instead of laying in the cells she roams over the combs, communicating her agitation to the whole colony. The bees fill themselves with honey before leaving; before the swarm rises, a few bees can be seen flying outside with their heads turned to the hive as though impatient to fly off.

At length a violent agitation begins in the hive and the bees appear, whirling in a continually enlarging circle like those made by a stone thrown into still water. Not a bee hesitates as it joins in the flight. The queen usually leaves after the wheel has formed and is airborne overhead with a roar like a tornado or a freight train. If the queen is heavy with eggs she may fall to the ground, unable to fly with her colony. The bees soon miss her and a search is made, the swarm scattering in all directions to alight on nearby trees and bushes. If they find her, the flight will continue with a solid core. Perhaps the bees have contrived a close flight formation with the queen aboard. If she is not found they return to the hive in a few minutes and another attempt will be made soon.

The queen sometimes alights first, especially if she is too heavy for a long flight; the bees follow. Again she will join the cluster after it has begun to form, but bees do not usually settle unless she is with them and she must be present in order for them to carry out the swarming plan.

After bees cluster, scouts leave to try to locate a new home. This will take from a few hours to several days and is the time for the beekeeper to put the bees into a hive. If the weather is hot and the sun shines directly on the cluster, they may leave before a suitable home is found, in which case they will cluster again in a more agreeable place.

Sometimes the queen, heavy with eggs, is unable to rise from the cluster for further flight. Then the bees attempt to build comb and found the colony where she is, on a post, a limb of a tree, or the side of a building.

Swarming bees will almost always alight where they see other bees clustered. Because of this, swarms of bees have been drawn to a selected spot by an old black hat or a mullein stalk colored black, which from a distance looks very much like a clustered swarm. A black sock or piece of cloth will serve well if fastened to a shady limb in the apiary from where the bees can be easily hived. Better still, a frame of old comb, dark from brood rearing, will attract bees and cause them to cluster. None of these is infallible, so it is wise to locate the apiary among trees and bushes if possible. If no trees or bushes are available and no settling place has been provided, the bees will settle wherever the queen alights, on a grapevine, on weeds, or even on the ground.

Although the bees in a swarm are peaceful, having filled themselves with honey, there are in every swarm a few from a neighboring colony or from the parent colony who failed to fill up on honey before leaving. They are likely to be angry when a swarm is taken and give the apiarist a rough time.

A swarm should be taken as soon as possible after the bees have clustered around their queen. Scouts may return later with news of a new home and the swarm will leave if the bees have not been taken. Most swarms that I have seen remained clustered for twenty-four to thirty-six hours.

When the scouts return, a few bees can be seen flying around the cluster. The number increases slowly until the entire swarm is on wing and away. The only way that we could stop a departing swarm was to hose them with a fine spray of water. Swarming bees are unnaturally heated and will welcome a cool and shaded hive.

A sure way to prevent a swarm from leaving their new home

is to put a frame of worker comb in the middle of the hive. Bees are so pleased when they find such unexpected treasure in a hive that they settle down to work at once. All of the swarms we took stayed in the hive; they were as pleased with a hive of frames filled with fragrant foundation as with a used comb. But giving too much comb will prevent the queen from following the builders, and too much drone comb will be built.

Swarms have often taken possession of a hive deserted by other bees and well furnished with comb, but they seldom enter empty hives of their own accord. I once found a swarm in a super containing combs of granulated honey that I had placed on a stand in the apiary for the bees to use. Since super combs are neither large enough nor sturdy enough to raise brood in, I had to transfer them to a hive containing wired foundation.

It is not wise to give a hive full of drawn combs to a swarm: the bees will fill the combs with honey at once, the queen will be deprived of enough room to lay, and the colony will decline in strength. If frames of foundation are used, the bees will bring in enough to build the combs and feed the brood. A very small colony with a good queen and some built worker brood combs will soon become a strong colony. In a poor honey season, built combs are a boon to even a large swarm, saving them the honey that building would require.

If a swarm alights on a small limb that can easily be cut with pruning shears without jarring the bees, they may be carried to a hive and gently shaken over it. All the bees will drop in; they quickly cover the frames of foundation and are in business.

One should never attempt to saw off a limb on which a swarm is clustered: the vibrations will anger the bees and you will not be able to hive them. If bees alight on a high limb that cannot be reached easily, they may be hived by the use of a swarm sack at the end of a pole. The sack is made of strong muslin and is similar to a butterfly net; it is about two feet deep, fastened to a wire hoop about one foot in diameter. The one I improvised from a pillowcase and a coat hanger was a good substitute; I imagine that most swarm sacks are improvised, because few apiarists have everything needed to hive a swarm under unexpected circumstances.

Care must be taken in emptying the swarm sack over the

hive, as this unceremonious imprisoning of the bees may anger them. If the queen can be obtained by trapping her as the swarm leaves, the swarm can be hived without any trouble. The parent hive is then removed from its stand and the new one placed there, with the queen caged on the alighting board. The swarm may cluster but will soon miss the queen and return; then the hiving is almost automatic except that you must release the queen and allow her to enter. The parent colony should then be returned to its stand, otherwise it will lose all the field bees, which with the loss of the swarm would cause it to be too greatly weakened.

When a colony has lost a swarm, enough bees remain to carry on the operations of the hive. The old queen leaves only when the population is teeming and there are thousands of young hatching daily; so in a short time the hive is almost as populous as before swarming. If the honey flow is great, the bees will be so busy that they will not be disturbed by swarming, especially as the older bees will be dying every day.

A new swarm contains bees of all ages, from those with the ragged wings of age to others so young they can barely fly. After swarming, bees that did not participate in the flight never attempt to join the new colony, and bees from the new colony never seek to return, even though the new colony may be near the parent hive. No one knows what prompts a bee suddenly to lose its love for the old home to the point of paying no attention to it.

If two swarms cluster together, they may be kept that way if abundant room for storing honey is given and one queen is removed. Large amounts of honey are obtained from such colonies if they issue early in the season. However, if more than two swarms cluster together, it is better to divide them. To do this, place three hives near each other and facing to form a triangle; deposit the bees on a cloth or paper between them. If most of the bees enter one hive, remove it a short distance, and try to find the queens and give one to each colony. If only one queen is found, it is best to cage her until most of the bees have entered. Then if they show signs of uneasiness and appear to intend to join the others, release the queen to them and all will be well. If one swarm is noticeably smaller than the others, it can be enlarged by shaking bees from the largest one into it.

If swarms that have clustered together are hived in a large box and left undisturbed until the following morning, they may be found in separate clusters in separate corners of the box. An apiarist once took five swarms that had clustered together and put them into a very large box and left them until fall. He found that they had lived together as independent colonies. Four were established in the four corners and the fifth in the middle, with a distinct interval between all colonies.

If two queens enter the same hive, they will often be found on its bottom board, each in a ball of angry bees. The balls must be dissolved and a queen given to each colony. When queens have been balled by mixed swarms it is well to keep them caged until the bees have quieted.

It is interesting to see how quickly a queen enters a hive when she recognizes the joyful note announcing that the colony has found a new home. She follows the direction of the moving mass, her long legs enabling her to outstrip her children. Other bees linger at the entrance, fly into the air, or collect in knots in front. But the fertile mother, seeming conscious of her importance, marches straight ahead into the hive.

Bees hived early in the day generally begin to range the fields in a few hours. If several rainy days follow the hiving, it is well to feed the bees to keep them from starving.

If an apiarist finds a swarm clustered nearby and is uncertain from which colony it came, he can take a spoonful of bees, dust them with flour, and keep them in a box until the swarm is hived. Upon release they will fly to the parent colony.

As many as three or four swarms may leave a colony in a short time, depending on hive conditions, weather, and honey flow. These are called after swarms; they seriously reduce the strength of the parent colony, since much of the brood left by the first queen has hatched and no more eggs are laid until swarming is over. In addition, if the remaining colony is too small to maintain adequate heat, great numbers of the coming brood may perish.

About a week after the new queen hatches, if the hive is too well filled and the honey flow is scant, another swarm may leave. This is called a secondary swarm—the apiarist usually tries to prevent it, as he sometimes does even with primary swarms. About a week after the first swarm has left, if he listens with an ear close to the hive in the morning or evening when the bees are usually quiet and hears the queen "piping," he can be sure there will be a secondary swarm, on the first or second day after the piping is heard. The apiarist can plan to prevent it: if weather is unfavorable, the oldest queen may be permitted to kill the others and swarming will be stopped. This is unusual, since young queens are not as particular about weather as are older ones, and may venture out even when rain is falling lightly.

The increase of colonies can be kept down if the after swarms are retuned to the parent colony. A swarm can be hived in any kind of a box and allowed to remain for twenty-four to forty-eight hours, at which time it is shaken in front of the hive from which it issued. Since there is a queen with them, a queen trap should be placed in front of the hive so she cannot enter. The bees will willingly enter their former home; seldom do they issue again.

As long as brood combs are fit they should never be discarded or destroyed, but used as part of every hive for a new swarm. Brood combs have been used for twenty years and the young bees were as large as the others in the apiary. Some combs have even been used for thirty years: they are good as long as a queen will lay in them.

When a swarm has left a colony the parent hive will contain a young queen, as the old one usually leaves with the swarm. A queen's life is not all that good, since according to this rule of swarming she must move every year of her life.

Recently I was told of the most fantastic place where a swarm of bees had alighted: in an old man's beard. He was a friend of her grandfather and had one of those old time foot-and-a-half-long beards. Peacefully sleeping in a reclining chair out in the sun, he was suddenly awakened by buzzing and the weight of bees on his chest. Fortunately he was not capable of sudden action or he might have been stung to death, but he did shout for help.

Members of his family came to find out what he wanted and when they saw the bees, they told him to remain absolutely still. One returned with large shears, the other with a box. The beard was carefully clipped off close to his chin and allowed to slip into the box, taking the bees with it. The old man was not stung but complained loud and long about the loss of his treasured beard.

It is a widespread opinion that bees are vicious and sting anyone who comes near. This is not true; bees never sting unless they are molested or cared for by an inept apiarist. However, there are bees in Kenya that the natives say are evil tempered and sting for the pleasure of giving pain. The natives don't care to risk the stings to get the honey.

North American bees and the Italian and German strains are mild by comparison, although we have had our troubles with the German variety. But these African bees will attack a person or animal two hundred yards away with no provocation. They also have a great capacity for swarming; the stronger the colonies, the meaner they are. If swarms should unite in flight, as often happens, they become increasingly ready to attack.

Their queens lay five thousand eggs a day, making it easy to develop strong colonies rapidly; they take over the hives of unwilling hosts. These bees migrated to Brazil in 1956 and spread rapidly over an area the size of the United States; and it is feared they may invade North America. A fund of $40,000 has been asked to study the problem and see what can be done to prevent their spreading north.

At first they lived in remote areas of Brazil, but recently they have been seen in Panama. Scientists fear they may be in the United States within ten years. There is even a possibility they might island-hop to southern Florida and proceed as our bees originally did. I wonder what the future of American apiculture will be if this should happen. It just might be that American beekeepers with modern know-how could handle them very well and our honey production would greatly increase.

Preventing Swarming

Swarming, as a natural method of propagation, a natural urge in bees, is difficult to prevent. It can develop unexpectedly in spite of our precautions against it, perhaps because our apiary is not near enough to our home for us to watch it closely. On occasion we have prevented primary swarming only to have a secondary swarm come out later to weaken the colony. When we were able to take the primary swarm there was no further problem of swarming during the season.

The method we used, making the bees imagine they have swarmed, is as effective as permitting a swarm to prevent further swarming. In our method, also called artificial swarming, Dave removed the hive containing the colony with queen cells—which means preparation for swarming—to a new location several yards away and set an empty brood chamber with seven frames filled with foundation on the old site. He then looked through the old colony until he found the queen and I took three frames of brood and bees from it, and put them in the new hive along with the queen. The field bees returned to the new hive, found their queen and very little brood or stores, and thought they had swarmed.

We then gave three frames of foundation to the old colony

and left them to make a queen, or if there was already a queen cell, a queen would hatch in a matter of a week. This colony would have a new queen, empty frames, and no field bees and they would think they had swarmed. Incidentally they are not queenless long enough to build an excess of drone cells.

We unite colonies only when one is too weak to survive the winter or has suddenly become queenless in the fall, and we use the newspaper method. If done correcly, the colonies will have united when small bits of paper are to be seen on the alighting board.

We did not try to force queens to live up to the rigid rules of the book; we were happy that they knew the rules of nature in their colony and that they lived by them. We tried to be scientific in our work but we were amatuer beekeepers, fascinated by our work.

Some colonies show no inclination to swarm, some start but respond to restrictions, and others swarm in spite of precautions. Our textbook suggested a couple of interesting ways to prevent swarming that we never tried. One is to take several frames of brood and bees from a strong colony, being sure there are eggs, larvae, and even a queen cell. Place the frames in a new hive, then take two frames of honey from another colony and add them together with five frames of foundation to the new hive. Move the old colony to a new site and put the new hive in its place so it will get the field bees. The new hive now has brood, food, field bees, and a queen cell. Replace the frames removed from the other colonies with frames of foundation. These last two are weakened and may not permit a swarm, the new one most certainly will not.

Another method is to take four frames of brood and their bees from a strong colony. The frames should have eggs, larvae, sealed brood, and emerging brood. Place the frames in a new hive; shake bees from two or three other frames into the hive and add five frames of foundation and move to another location. All field bees will return to the parent hive, so be sure the queen is in there, and the new colony will make a queen for themselves.

When one of our colonies showed signs of swarming Dave decided he would prevent it, so he set a queen trap in front of the hive. The bees opened a hole on the corner of the hive and left with their queen; all Dave had for his trouble was a trap full of fat drones.

He said, "Oh, if bees were only swarmless!"

In fact, swarmless bees used to be produced by breeding, but they were lazy, unprolific, and poor gatherers of nectar. Beekeepers must take bees as they are if they want honey.

One noon I noticed a number of bees in the living room and wondered if there was a hole in a screen or a window open somewhere. The next morning the room was alive with bees. I went outside and saw a line of bees leading from a clump on a bush, like ants on the ground, to the top of the chimney. This was unusual because bees neither walk on the ground nor move in a column. I put some excelsior in the fireplace and lighted it to make smoke; then I went outside, cut the branch from the bush, and shook it over an empty hive. When I took it to the living room, the hundreds of bees flying around were in it in a short while and I put them out in the backyard.

One spring Dave looked into a hive that had wintered in the backyard and called in alarm. "Wow! There are nine queen cells in here and one is capped. They may swarm anytime."

"They certainly could," I said. "The old queen could leave any minute and the first of the nine to hatch would take over the hive."

"I'll get a hive ready." Dave went to the garage.

Swarms usually issue before noon but none came out. Three days passed and no swarm, so Dave decided to have another look. "Way out!" he exclaimed. "All the queen cells have been torn out but the bees are making new ones. Do you suppose they are fooling us?"

Later on when we were not expecting it they swarmed, but we did not get them. They whirled out, mounted high, and flew away. Bees can always surprise you no matter how long you work with them.

If too many swarms are lost the beekeeper is unable to build his colonies, so prevention of natural swarming is an important item in his work, even if he has as many colonies as he cares to keep. Many apiarists, especially farmers, keep bees solely for honey; since in many seasons it is not possible to increase their colonies to get a good harvest of honey, they choose to restrict swarming.

Frames in hive or super

a and a: side view of frame - top of frames, in hive or super

b: walls of hive

b and b: side of hive or super

Another objection to natural swarming arises from the fact that bees are liable to swarm too often and weaken the parent stock. Beekeepers remedy this by hiving the first good swarm and returning all others to the parent colony.

Again, bees may be located near a highway where cars and people pass constantly: accidents may occur. Or they may be near a woods where swarms cluster high on a limb and it is impossible to hive them.

Then too, it is very annoying to have to watch the bees for weeks hoping to get the swarm, or to have them swarm at unexpected or unwelcome times when the owner is engaged in other business; many beekeepers are also doctors, lawyers, or merchants with daily duties.

The many swarms lost each year are a further argument against natural swarming. An eminent apiarist has estimated that one-fourth of the best swarms are lost each year.

Although swarming is a natural impulse in bees, it can be prevented if it is caused by abnormal conditions in the hive—usually, lack of room in the combs for the queen to deposit her eggs and for workers to store honey. The addition of a second brood chamber or causing artificial swarming can remedy this.

Not every colony allows a swarm every year, only strong ones with large populations, good stores, and crowded quarters. Colonies are only temporarily reduced by swarming: they have the power of replacement within themselves. Before swarming, the bees' attitude toward the queen changes; they no longer attend her carefully and often refuse to feed her. She wanders over the combs, her egg laying diminishes, and her abdomen shrinks so she will be able to make the swarming flight.

Only beekeepers realize how suddenly a honey flow comes and how rapidly the combs can be filled, causing the desire to swarm. Strong colonies, almost destitute of supplies due to the large amount of brood being raised, have been known to bring in twenty pounds of nectar in a day. Conditions like this, with an increasing population, cause swarming. Even placing supers for honey will not suppress the urge unless the weather turns bad or the honey flow stops.

Dividing a large colony into several nuclei has not always solved the problem, for each nucleus may swarm if "swarming fever" strikes. The heat of summer hastens preparations as does the appearance of excess drones in the brood chamber. These big, lively fellows make the already crowded hive more uncomfortable.

Hives should be located where there will be shade during the hottest part of the day, either in a beehouse or in the shade of trees.

The easiest, and most commonly used way to thwart swarming is to divide a colony and add five frames of foundation to each half, then let the bees go on from there. One half will have the queen and the other half will create one, so no lengthy examination of frames or searching for the queen is required. This method should cause no unusual loss of bees; for even a large colony diminishes for several weeks after being hived. So great is the loss of bees during the height of the honey flow that at its close colonies may contain only half of their original population. They rebuild in several weeks in order to have a strong colony for the rigors of winter.

Populous colonies that are building queen cells during the early part of the honey flow are apt to let out a swarm when the queen cells hatch; the queen cells should be removed to prevent this. One new colony made before the honey flow is worth two made later, so all swarming prevention should be done early. This will result in more honey and less loss of bees, and will mean you don't have to watch for weeks, only to have a swarm come out at an unwelcome time.

An Occupied Attic

One Monday after Dave had gone to school and I was hanging laundry on the line, bees kept bumbling around in my hair and alighting on the damp clothes. I looked up to see where they were coming from and to my amazement it was through the louver in the gable of our house. It usually has a window back of it but in the summer the window is removed for ventilation. Could there really be a swarm in the attic, I wondered, and if there were, when did it get there? It had to be sometime when Dave and I were both away from home or we would have heard it come. Many times I have seen swarms fly across the yard at less than treetop height on the way to someplace the scouts had found, but they had never entered our house.

When Dave came home I said, "Guess what? I think we have bees in our attic."

"Zowie! You're kidding."

"Indeed not, come and see." I showed him the bees coming through the louver.

He whistled and said, "Way out! Maybe we should wait until after sundown to take them, since we don't know how long they've been there and there will be field bees coming in."

"That's what I thought. Besides, we'll have to have a light to see to take them anytime, so it might as well be this evening."

After dinner Dave put the stepladder in the hall below the

entrance to the attic, climbed up to remove the trapdoor, and set the Coleman lantern on the attic floor. He got a hive body with five frames of foundation, a cover, and a bottom board, and took them up. Then we went up in shirts, jeans, boots, veils, and gloves. We knew the bees would be angry, so we were prepared.

It was certainly a sight to behold in the mellow light of the lantern. The bees must have been there for some time, for they had built several combs and fastened them to a rafter. They looked like the petals of a huge water lily pinned upside down to the rafter; several petals were long and some a bit shorter, with bees all in between them.

Dave was five feet eleven inches tall and able to reach without a stool the place where the combs were anchored. With the big knife he carefully cut the petals loose and put them between the frames. He knew the queen would be there even though bees were flying around; she would stay on a petal of the comb just as she stayed on a frame when we inspected a hive at the apiary. The bees in the air were very angry and assailed us in the semidarkness of the attic.

Dave smoked the rafter where the combs had hung and we left the hive stand uncovered until the next evening so the bees we had disturbed would settle in it with the others. After dinner we went up to bring them down and all was quiet. Dave put the cover on, stuffed the entrance, and tied a rope securely around it to take it down the ladder. We discovered that I had to stand on the ladder while Dave carefully lowered the hive to me and then I had to back down with it. Dave took it out to the stand in the backyard, where it remained for three weeks until all the brood had hatched from the comb petals.

Dave would check it every few days and remove any of the comb that was empty of brood. When at last all the petals had been removed and replaced with frames and all the bees shaken into the hive, he checked to find the queen.

"Is she ever beautiful!" he said. "All fuzzy and golden and she is a good laying one too. Some of the frames have large circles of brood on them already." We took the colony to the apiary on our next trip; but before that Dave tacked some screen wire behind the louvers, so we would not again be unwilling hosts to a swarm of bees. All the same, it was one of the easiest bee takes we ever made.

Sweets to Japan

One fall my daughter Eileen was in Japan for the school term. Her husband, a doctor of physics, had been asked over to teach in one of their universities. She wrote me saying: "Japanese honey is very different from American honey, and while we use it, we long for some of yours."

"Why don't we send her some?" Dave asked eagerly. "I bet it would really surprise her."

"I'm not sure we could," I said. "There may be some sort of postal regulation against it. You will have to call the post office and get information on how to send it."

Dave made the call and after some time came to the kitchen with shining eyes and some notes.

"We can do it if we put it in a tin, seal it securely, and mark it 'confections.' "

In the fruit room I found a one-pound coffee can with a lid; it would hold three pounds of comb honey, which was the kind Eileen wanted. There was a box of paraffin on a shelf, left over from jelly making, so we took it to seal the can and prevent leakage.

I selected a nice full comb from one of the supers, and cut it from its frame. Using the bottom of the can as a guide, I cut three round pieces from it. These rounds completely filled the can, as it was three inches tall and the comb was one inch thick.

I put the lid on and said, "It's going to work nicely, now melt the paraffin so we can seal it."

"Why can't we use beeswax instead of paraffin?" Dave asked.

"Why I never thought of that—you are a beekeeper at heart. Of course we can use beeswax, we have a lot and it certainly sticks to whatever it gets on."

Dave melted the beeswax and I held the can sideways over some paper and rotated it as he poured the wax along where the lid overlapped the can. We let it cool and repeated the operation several times until the can was tightly sealed. I wrapped it in corrugated paper, then brown wrapping paper secured with gummed tape, added an address label, and it was ready for its journey. Dave printed the word "confections" on the top, bottom, and sides and we went to mail it.

When it was inspected, weighed, and insured, Dave grinned and I said, "I hope it arrives safely and doesn't leak and stick to everything."

"Why don't we send it airmail so it won't take long to get there?" Dave asked.

We went home feeling that our mission had been accomplished, and waited for word from Eileen. In about ten days we had our answer.

"What a nice surprise," she wrote. "Why did you ever decide to send comb honey? It was in perfect condition; not one little drop had leaked out. I never once thought you would send me some. Honestly, I wasn't hinting when I told you about the Japanese honey. All my friends use it and think it very good, so I guess it depends on what one is used to. However, the same friends thought mine was superb when I served hot biscuits with thin slices of comb honey on them."

Dave was delighted with the outcome of our venture and a year later, when Eileen was back in Santa Barbara, he wanted to send honey again.

"Since it got to Japan safely, it should travel to Santa Barbara OK, shouldn't it?"

To this day I wonder whatever went wrong; for, like Dave, I thought that since we had sent a can of honey overseas successfully we should certainly be able to send it to the West Coast. But when Eileen's letter came and we opened it expectantly, there was bad news.

"The postman rang my doorbell, and holding a sticky package up before me, asked, 'Do you want this thing, whatever it is? Or do want to file an insurance claim on it? I think it is broken, whatever it is!'

"I took it and opened it in the kitchen sink. There were cracked places in the wax sealing it. I guess it was handled more roughly than the airmail to Japan. But most of the honey was still in the comb and we are enjoying it very much. Maybe you shouldn't send any more honey or Uncle Sam's mail will be all stuck up."

Dave was convulsed with laughter over the picture Eileen's letter had conjured up. "I guess I didn't get it sealed as well as you did. Don't ever tell her I was the one who wrapped it."

The Usurper

Some of the swarms we took were the smaller, very dark German bees. We found them quite short-tempered, not so much when we took them but when we checked them later and when we harvested their honey in the fall. If we were so unfortunate as to kill a bee accidentally we had to leave our work in the apiary and return later, for one angry colony can set others on the warpath. Dave was unhappy about it and made a suggestion one day.

"Why don't we get an Italian queen for them and see if that helps?"

"It will help by fall when we harvest the honey, because all of the present colony will be dead then and we will have a nice mild colony. Right now it won't help much."

"Let's try ordering again," Dave suggested. "Maybe we'll have better luck this time."

Queens can be ordered from bee raisers and will arrive in about three days by airmail. So the second day after Dave had sent the order we went to the apiary to remove the German queen, for a colony must be queenless for at least twenty-four hours in order to accept a new queen. This was rather hectic as all the bees objected and were quick to make it known.

In due time the tiny package arrived and Dave was pleased. He set an empty hive beside the occupied one and removed the cover while I ran the smoker, first a puff at the entrance, then under the cover. He began to examine each frame, looking for the queen.

"Zowie! The bees all run around so fast, this isn't going to be easy." But he took his time; after examining a frame he gave it to me to put in the empty hive. If we returned it to the colony the queen might run back on to it and we would never find her.

The outer two frames on both sides usually contain only honey, but Dave examined all of them in order, starting at one side. The queen was not on the first or second one nor even on the fourth or fifth, but as he picked out the sixth frame and looked at it, he said, "She's on this one, but she runs so fast—and so do the bees—that I can't keep her in sight. Can you do something?"

"Just hold it a minute and I'll spread a sheet." I spread a small sheet in front of the hive. "Now shake the frame over it and if she falls I'll pick her up."

He shook the frame gently and many bees fell off, but no queen. He shook it a couple more times while I watched, still no queen. By this time there were very few bees on the frame; as Dave turned it over he saw the queen, surrounded by a cluster of bees, her caretakers. I took a small box from our supplies and swept the cluster into it and we had the queen.

Dave replaced the frames in the hive, covered it, and we left the bees to moan the loss of their queen until the next day when we took the new queen out and hung her in the hive, in her little cage sealed with a plug of hard sugar. If all went well the bees would eat the sugar and release her into the brood chamber.

"I sure hope it works this time," Dave said. "If it does, we'll have a nice mild Italian colony before too long."

But it didn't work. When we went out in forty-eight hours we found the cage empty and the queen dead on the floorboard.

"I can't understand it. That never should have happened," I said. "We know they were queenless and there were no queen cells in the hive." I thought a minute. "Or did you look?"

"I didn't see any on the frames I looked at. Do you suppose they didn't want the Italian bees that were with her?"

"That is possible; however, bees are always sent with a queen or she would die on the trip."

"At least they got her out of the cage. I wonder what happened then," Dave said.

"They must have balled her—but why, we'll never know."

"What do we do now? We still have the German queen at home with her little cluster of bees that are feeding her the honey I put in with them. Should we bring her back?"

"But we wanted to convert the colony to Italian bees, and if we bring the German queen back, there goes our chance," I said.

"They'll make another queen if we don't do something."

We are always prepared to open a hive unexpectedly, so I said, "We'd better open several hives that have good Italian stock until we find a sealed queen cell and give that frame to our problem colony. It shouldn't be difficult, since it is spring."

Dave began the search, which was much easier than trying to find the German queen had been. He removed a frame near the center of the hive where the brood would be; it contained no queen cell, but the next one did. He brushed all the bees back into the hive and put it in the German colony, from which I had removed a frame near the center. Dave brushed all the German bees from it and gave it to the Italian colony.

He laughed, "That was sneaky, a double switch if there ever was one. Now there will be a frame of German bees in the Italian colony and a frame of Italian bees in the German colony, plus a nice Italian queen when she hatches out. I hope it all goes well."

In a week we went to check the hive and Dave opened the German colony to find the queen. She was easy to find, being larger and lighter colored than the German bees.

"I wonder why re-queening always works when we give a queen cell, but never does when we buy a queen," I said.

"I guess it is because she hatches from a comb in the hive and bees naturally accept whatever hatches there," said Dave.

By midsummer our angry German colony was a thing of the past and in the fall, harvesting honey was much easier without an angry colony to alarm the others. A much quieter atmosphere pervaded the entire apiary and we were happy about it.

Hunting Bees

I first heard of hunting bees one day when I was at the apiary to take a swarm from one of our fence posts. The swarm had issued from one of our colonies and I was placing an empty hive body by the post to transfer the swarm to it when I became aware that I had company. Otto, an elderly beekeeper from across the road, had come to tell me how to do it.

With the bee brush I carefully brushed part of the cluster onto a frame of foundation and placed it in the hive. When I had filled four frames I had the bees in the hive except for a few that had fallen on the ground and some flying around. I put the cover on at an angle, smoked the post with the smoker, and placed a small piece of wood from the ground going up to the entrance so the bees on the ground would walk up into the hive. Then I turned to Otto; he had been talking all the while, but a beekeeper taking a swarm must concentrate on the situation, so I had not listened attentively.

"Good work!" he said of the final result I guess, for I was sure I had heard remonstrances as I worked.

We sat in the shade of a tree and Otto asked, "Have you ever hunted bees?"

"Can't say as I have. All that I come in contact with are thrust upon me. Hunted bees?" I questioned. "How do you hunt bees?"

"I've hunted hundreds of 'em, I guess."

"Where do you go to hunt them?" I asked. "Do tell me about it. It sounds interesting. You see I am usually summoned by some alarmed individual who has bees she doesn't want and is afraid of, unless it is a swarm of my own, like this one."

"First I put a hive box, maybe two, a rope, an ax, a ladder, a cloth funnel, and a bee escape in the car. Oh, yes, I'll need a hammer and some nails too."

"What's a cloth funnel?" I asked.

"It isn't really cloth; it's some kind of woven wire like screen wire. Then I drive into the hills until I come to a meadow with trees all around it. I have fixed a bee trap out of a cigar box and a piece of glass and I always take that along.

"Bee hunting is a real pastime like fishing, so I take a whole day for it. Of course if you want bees in a hurry, you better buy them or take them like you just did." He motioned to my swarm, which was pretty well in the hive by now.

"Yes, I like to take them in as short a time as possible."

"Oh, it's good that way sometimes, say for instance if there is foul brood in a swarm in a tree. Then when I locate the tree and take the swarm, I'm helping every beekeeper in the area."

"You certainly would be. But what about the honey left in the tree? Bees would rob that and take it into their brood chamber. Infected honey is what starts foul brood in a colony."

"I take care of that too. I seal the hole up so no other bees can get to the honey. As I said, I have this cigar-box trap. I made a hole an inch in diameter in the lid and fixed a piece of sliding tin over it so I can open or close it as I want to. In the bottom of the cigar box I put a saucer with honey diluted with water to about the consistency of nectar. I float a paper on it, like we used to do with that old black, poison flypaper."

"That's so the bees can't drown in the syrup, isn't it?" I commented.

"Right!" Otto said. "Something else I'll need is field glasses to see where the bees go. I go to the meadow where flowers are in bloom and look for bees gathering nectar. I don't bother with those who are gathering only pollen. I trap a bee in my cigar box by holding the box under her and shutting the lid on flower and all; then I slide the little tin lid over the hole. I leave her in there long enough for her to fill up on the honey water, then I put the box on a stump, a fence post, or my ladder, and open the tin slide. The bee, filled with honey water, comes out into the top box with the glass window. I draw back this glass slide and let her go and step back and watch her. She will circle around the box so she will know how to come back, then the circles grow bigger and bigger and she flies off in the direction of the bee tree."

"Why do you feed the bee? Why don't you just trap her and let her go and watch her?" I asked.

"Some beekeeper you are." Otto laughed, slapping his knee. "You should know that unless a bee is loaded with nectar she won't go home. If I didn't feed her she would just go to another flower and I would lose track of her. Well, she'll be back in a few minutes, the time tells the distance. It takes a bee about eight minutes to make a round trip of half a mile, and thirteen or fourteen minutes for a full mile. This allows for her to unload and return and for the circling around as well. While she's gone I trap another bee, feed her, and let her go. If all the bees I trap go off in the same direction, I pick a point in the woods in that direction."

"It must take a long time to locate the tree. I had no idea that hunting bees was so intricate."

"It sure is. You wouldn't do it, but you see I'm out for an afternoon's sport just like a fisherman. Sometimes the bees go off in different directions and that means there is more than one bee tree around. Soon the first bee returns and the box is open and ready for her. On her second or maybe her third trip she may bring other bees with her—this makes it easier to follow them. I check the time and right off I know how far it is. When bees are going and coming pretty rapidly, I move the box nearer but on the same line of flight to the tree. I watch only the bees that fly in that direction. Then I move the box to one side and line things up as before to establish a cross line of flight; where the two lines meet is somewhere near the tree."

"Bee hunting is quite scientific, isn't it? When you first mentioned it I had an idea it was a real haphazard thing," I commented. "I thought you just went out and looked for bees in dead trees."

"Bees seldom go for dead trees. Maybe they know they might

be blown down or that rain might soak through the porous wood and they would get wet. Now I take the field glasses and look in the direction of their flight for holes or hollows in trees. If I don't see anything I make another cross line off to the other side and where the three lines meet just has to be the tree. I look through the glasses again and most of the time I find the tree. I go and carve my initials on it. This makes the swarm mine but doesn't give me the right to cut the tree to take them. For that I must get permission from the owner—which he usually gives if I promise some honey. However, I seldom cut trees. I climb and get them out if they are in a hollow tree."

"So even bee hunting has to be legal," I said in surprise.

"It sure does. After I initial the tree, maybe I'll go back and take up another flight line and follow it to another tree. I can always take my time getting the bees after I have initialed the tree."

"I never would have thought it had to be done legally or that someone else could claim the bees. I didn't think people bothered about bees in trees in the country."

"It doesn't happen very often, someone claiming the bees, but it has been known to happen." Otto was silent for a moment and I started to collect my tools to go, but he continued.

"Oh, yes, if a limb is to be cut or a tree chopped down, you sure can't start without blowing plenty of smoke in to quiet the bees, so I always have my smoker too. I don't like to cut trees down; but as a rule a tree with a hole in it is of little value to the owner for lumber, so he gives permission. After it is cut down I can chop into the cavity and get both the bees and the honey. When the tree crashes down the bees come rushing out and I sure need the veil and smoker then. After a few blows with the ax the bees quiet down and the smoker will do the rest".

"Suppose the bees are high up in the tree and the owner doesn't want you to cut it down, what then?" I asked. I was impressed by Otto, who had good answers to all my questions and never lost the thread of his story.

"I just forget it. The risk of climbing is more than the bees are worth. Actually I have cut very few trees: the bees are usually where I can get them with my extension ladder. This is true of a building where they have made their home, too.

"To get the bees I use the wire cloth funnel or cone and put a frame of brood with a queen cell in the hive box. I use the hammer, nails, and lumber to make a stand or platform to hold the hive, then I move the ladder up to the tree until the hive is directly opposite to it and no farther away than the length of the cone from the hole. If I want the queen, I just tack the cone over the hole; if I don't want her, I tack a bee escape under the cone. Either way the bees can come out but can't go back, so they will go into the hive with the brood and honey."

"How long do you leave the hive there on the ladder or platform?"

"At least three weeks, so I get all the brood and honey. They will move it out too when I remove the bee escape and the cone. I do this when the swarm is well located in the hive, queen and all, if I have used just the wire cone. If I've used the bee escape, the queen has hatched from the cell in the hive. Now I load the smoker and brimstone and what is left of the old colony, not more than a handful of bees and a queen if I used the bee escape. When I leave I don't replace the cone or bee escape, and what do you think happens?"

"I really have no idea. I don't suppose the bees would go back in or you wouldn't leave the cone and bee escape off. What does happen?"

"You're right. They won't go back into the brimstoned hole. I just leave them there for about three days; then I go back, take down the stand and ladder, and take the bees home. Before I go I close the hole with a plug so no more bees can occupy it; for they would be attracted to it when the brimstone odor clears out. So you see, that way no damage is done to the tree.

"I take bees the same way if they are in the side of a house, but the trouble there is that people don't want a hive around long enough for me to get the bees, let alone the brood. People want it done instantly, so I tell them to call an exterminator, because I don't want to kill bees. I have taken bees from old vacant houses that way, though."

"I had to kill bees twice, once in a chimney and once by a porch, but I will never do it again." I arose to go, as my swarm was quiet in the hive now. "I still prefer my way of taking swarms, but I'm so glad you told how you hunt bees."

We said goodbye and I returned home.

Another bee hunter said he hunted bees only for their honey

and didn't care what happened to the bees. "I saw or chop the tree down. The crash of the tree coming down paralyzes the bees; then I make a cut below the entrance and the hollow in the tree is exposed. The honey can then be removed. Why, I've gotten anywhere from a pailful to ninety-seven pounds of honeycomb. I do this early in the spring and the bees will have time to replenish their honey for winter. And if they don't—"

Quite a difference in the concern of the two hunters.

Reverting to Type

Not long after, I was forcefully reminded of Otto's bee hunting, only I did not have to hunt these bees: they settled on a branch of an elm tree in our backyard, about thirty feet up and ten feet out from the trunk. I was in the yard when they came and heard their humming during flight; then I saw them begin to clump on the branch.

When Dave came home from school I said, "Guess what we have in a tree out back."

"Not bees again, I hope." He took his books to his room and returned. "Where are they? Let's take them."

We went outside and I showed him where they were. He whistled.

"We can't take them, they are too high," I said.

"No they aren't. You took some from the other tree."

"The others were not that far up, only about twenty feet. These must be thirty or more."

"That doesn't matter. I can climb the tree and work my way out along the branch and shake them off or even brush them off if I can get that close to them."

"You could neither shake nor brush them off because if you did, they would fall to the ground and we'd really have a problem. They would be killed or mad."

"Couldn't you stretch a sheet or something to catch them?" he asked as he began to climb the tree.

I went and put my arms around his waist. "Just a minute! Let's talk about this. That branch you are planning to go out on isn't very big. It won't hold your weight. Just see how it bends under the weight of bees. You would be what would fall and a sheet couldn't catch you."

He stood looking up and thinking. "You're right."

Dave weighs 135 pounds and he realized that was quite different from five or six pounds of bees, and they were bending the branch considerably.

"They will probably be gone in a couple of days," I said.

Every morning and afternoon Dave would go out to see if they had gone, but they hadn't. I too began to look up at them every time I went out. After three weeks they were still there and they were making comb, petal-shaped pieces like the ones in our attic. These were larger and fastened to the branch; they looked something like combs cut from frames, and I wondered if the man who had first devised frames had seen combs attached to a tree like that.

The bees were still there by frost, which was what we expected when we saw they were making combs. As the leaves began to fall, pieces of comb came with them until when the tree was bare, only a small piece of comb was still fastened to the branch. We saw no bees or brood on the comb that fell, nor was there any honey, so we wondered if they had flown away one mild morning before frost. If they did they soon died, because they would have no stores for winter. I have always been sorry about that swarm but there was no way we could have taken them. I guess that is nature in the raw: the bees were reverting to type and very unsuccessfully.

A Free-flying Plane

This chapter is in a book on beekeeping because Dave made the airplane and the fire department eventually got into the picture. The plane was twenty-six inches long with a wing-spread of fifty inches. Its tail was twenty-four inches long and five inches wide and had no fin; the wing was five-and-one-half inches wide.

Dave made the frames for body, wing, and tail from balsa wood and covered them with silkspan, a very fine paper for making model planes. He painted them with model airplane dope until the silkspan was like fine vellum. The plane was a long time in the making, as he worked on it in his spare time, and what with school and beekeeping he was quite busy. It had a small gasoline "baby bee" motor connected to the propeller and a small gas tank underneath. It was really beautiful when it was finished.

"Let's go to the park and fly it," Dave said the Saturday after it was finished. "I want to see if it really will fly alone."

There were many people in the park, mostly children. Dave started the motor and tossed the plane into the air. At first we thought it was going to crash, but it mounted slowly and flew in great circles overhead. The circles widened and Dave became very excited, then: "Oh, no! It's going to crash into that tree the next time around."

He was right. The tree was very tall and the plane settled on a branch up near the top like a great white bird.

"Well, that's that," I said. "I see no way to get it down."

"No, it isn't that! I'm not going to go away and leave my plane up there after all the work I put into it. I'm going to wait and see if it will fall or blow down. You can go home if you want to."

"Dave, you can't wait here. There isn't a breath of air; it might take days to come down; it might stay there like a kite in a tree."

But he was determined, so we waited for a while. We had quite a gallery by now that had assembled as the plane flew, and everyone was talking to each other or to us, suggesting what to do.

One small boy said, "I'll stay here till it comes down and get it for you if you'll give me a dollar."

"No you won't," his mother said. "You're not going to stay here overnight to get someone's plane. That thing won't come down till there's a wind, and no one knows when that will be." Someone else offered to climb to the plane and toss it down. "You're crazy," said the youth with him. "A cat couldn't get up there."

"Let's go to the fire department at the far end of the park and see if they will help," I said to Dave.

"Not a chance," he answered. "They don't bother with planes caught in trees, and if they did it would cost more than the plane is worth."

"Ordinarily that is true, but maybe this will be different."

"What makes you think this will be different? Do you know someone there?"

"Maybe I can pull some strings." I smiled, so we drove over. There were calls from our gallery, who thought we were leaving.

"Can I have it when it comes down?" "No, it's mine. I offered first." That was all we heard.

I introduced myself and Dave to the desk sergeant and said, "We are beekeepers and are on your bee pickup list. Dave's

Dave's plane in tree

free-flying plane is caught in a tree in the park and I wondered if you would be so kind as to help him get it."

The sergeant looked his amazement and started to object.

"We must have answered over two dozen calls for you, otherwise I wouldn't have thought of asking for your help. I saw the hook and ladder car parked just outside and the tree isn't far away."

He huffed a bit, then called to a man in the other room: "Orville, you got that hook and ladder truck all checked over now?"

A fireman appeared in the door drying his hands on a towel. "Yes, sir."

"You going to take it for a trial run soon?"

"Yes, sir."

"Take it now and take Dave here, to get his plane out of a tree. He'll show you where."

"Yes, sir."

Dave got into the truck with the fireman and I followed, after thanking the sergeant. Orville sent the ladder up until it was directly under the plane and said to Dave, "You go up after it."

The gallery was still there and had quadrupled as the fire truck came. Dave climbed the ladder and got his plane. Hands reached for him and the plane and he had to hold it over his head to keep it from being destroyed.

Someone asked, "How come you got the fire department to get it for you?" Another asked, "You a rich guy or somethin'?" A mother asked me, "What did it cost you?"

Orville winked at Dave, retracted the ladder as we thanked him, and drove off. We too drove home, leaving all questions unanswered. On the way Dave almost hugged his plane and said, "I almost flipped over what you did. You were the greatest—thanks a million."

Neighborhood Problems

The only local bee problem we had was the one when Dave had to kill bees because they were by a neighbor's porch. The year after Dave left for college, the same neighbor had another problem.

A large swarm flew past my kitchen window as I was washing my breakfast dishes one morning. I thought they had settled in a large spruce tree in the neighboring yard, as they were gone suddenly, but they lit on the second-story window of Jim's apartment house. I had never seen bees cluster on glass, but they were draped over the window like a curtain, about twenty feet above the ground.

The next day when I returned from church, Jim came over and told me about them.

"They're on the bathroom window. They came yesterday—can you do something about them?"

"If you have a tall ladder and I can get someone to help me, I could do it. That's a job for two people, one on the ladder and one on the ground. Since Dave is gone, there is no one to help me."

"I have a ladder, but I couldn't help you," Jim said. "And I wouldn't want you to go up. You might fall."

I began to wonder how he thought I could take them without climbing the ladder. Maybe he thought I could charm them off the window. It was a large swarm and covered the upper corner of the window.

When bees walk on a polished surface like glass, the climbing claws are folded back and a pocket on the bottom of the foot secretes a sticky substance that enables them to cling to the smooth surface. The claws cling to objects like trees or a post when bees climb or hang in a cluster. These bees were clinging to glass, to each other, and some above to the window frame.

Jim had been thinking. "No, I don't want anyone going up a ladder, they might fall, get hurt, and sue me. Couldn't you just kill them like you did the others?"

"From down on the ground? No indeed, I'd have to be on a ladder even to kill them. Don't you remember, Dave was on a ladder the other time?"

Since he would not consent to my using a ladder, that relieved me of the task of killing the bees without an outright refusal, which might have made him angry.

"Since they came yesterday, they should be gone by tomorrow, or Tuesday at the latest. Why don't you forget about them? They are up on the side of the house away from all doors and will not bother anyone. They will leave when the scouts find a permanent place, and that usually takes several days," I told him.

"But they are bothersome. I found several in the hall upstairs, and Sally—she's the occupant of that apartment—is afraid of them."

"Have you examined the window for a crack where they are going in? If you find one, stuff it with cloth or paper."

"I did find a place and I already stuffed it with cloth."

"Have any more gotten into the house since then?" I asked.

"No, but I want them off the window at once!"

"Just be patient, they will leave in a day or so. And since I have no one to help me and you don't want me to climb the ladder, there is really nothing I can do. I'm sorry," and I turned to go into the house.

"I'm going to call an exterminator first thing in the morning! I don't intend to wait for them to go away!" His tone sounded like he was more disgusted with me than with the bees.

Jim called to me on Monday morning when I was out in the yard, "I got the exterminator. He'll be out here by noon to kill them."

As we talked, another swarm lit on the adjoining window in the same fashion and Jim was appalled. "Why I never heard of such a thing! Another swarm. Oh, well, the exterminator can kill them both when he comes."

The last swarm covered Sally's bedroom window and I hoped there was no crack around it where bees could creep in. Swarms often light near another swarm out at the apiary if they come out at nearly the same time, but two swarms on adjoining windows of a building in the city was fantastic.

The exterminator had not come by late afternoon and I began to wonder if he intended to come, since they do not like to kill bees, especially out in the open. They know the value of bees in pollinating trees, flowers, and crops; and then too, many preparations are not effective on bees. We found this to be true when we took a swarm that a woman had been spraying with an aerosol fly killer. The bees were lethargic and easy to take but they recovered in the hive the same day.

Later, I was watering my rose garden and Jim called. "The exterminator just called me and said he couldn't come this morning because he had a tankful of weed killer in his equipment and would have to use that first, but he should be here by evening."

"They could be gone then, I hope," I said.

By night the swarms were still there and the exterminator had not come. On Tuesday morning there still was no change, but as I was doing my breakfast dishes I saw activity among the bees. In a moment the air was full of them, whirling off somewhere. I hoped it was both swarms, but when the activity was over the second swarm was still on Sally's bedroom window. I had a luncheon engagement and when I returned in midafternoon, one glance told me that the exterminator had been there, for piles of dead bees were on the windowsill and the large swarm was gone, dead on the ground.

Jim saw me and called, "The exterminator came after you left and now I'm rid of bees."

"So I see," was all I could say.

"It was just as well that he came while you were away,

because you love bees and you wouldn't have liked it."

At least he had some consideration for me. "Yes, I love bees, I love all nature and people as well," I said. I was sorry he had not waited a bit longer and the second swarm would have left too. It had cost him twenty dollars to have the one swarm killed but perhaps he did not object to that.

A friend at the north end of the block called me one day and asked if I would come and see what was living in one of her fruit trees. She met me at the door and we went to look at the tree. It had an immense trunk for a fruit tree, but about four feet above the ground it branched into three arms. Right above this division there was a small elongated hole in one of the arms.

"Look at those insects coming out. What are they?" she asked.

"Bees," I said. "But I can't imagine where they are living; it must be a small swarm, because the limbs are small, unless they go down into the trunk."

"Can you get them?"

"I could, but I don't think I should try. I would have to set up a hive and trap them," and I described a process somewhat like the one Otto follows in bee hunting. "The hive would have to be there for several weeks to get all of them and since you have two small children it might be disturbed or moved, and that could be dangerous."

"What can I do? They can't stay there; the children might be stung or even Don or I."

"The best thing to do is to call an exterminator. He will kill them in a matter of minutes and you will be free of them."

"Could you do it?"

"Sorry, I don't keep on hand anything to kill bees, but he does." She followed my advice and we remain friends.

Last spring I was returning from a store around the block and saw a small black kitten playing with something on a lawn under a bush. As I came closer I saw that it was slapping at a swarm of bees on the ground. As I watched, a woman came out of the house and called the little kitten.

"I'm afraid she'll get stung," she said as she picked the kitten up. "The bees came yesterday and were on the bush but a boy who lives across the street threw a coat hanger at them and they all fell to the ground. Do you suppose you could get them up?"

"I doubt it, but I can try," I said.

"Oh, that would be fine, because I don't know what to do about it and some of the children might be stung with them there in the grass. There used to be a nun who kept bees and I tried to call her yesterday when the bees came, but was told that she doesn't do that anymore. So I would be glad if you would take them."

"I have never taken bees from a lawn and it might not work. It reminds me of trying to get sugar up out of sand, but I'll try."

I went home and returned with a hive, a few frames, and a cover and bottom board and set up under the bush and near the bees. I carefully shoved a thin piece of plywood as close as I could to the bees and slanted it up to the alighting board of the hive, hoping they would crawl up, and I left them that way.

Just then the woman came out of the house and said, "It looks like we might have rain—look at all those clouds. What will the bees do then?"

"Nothing much, unless it is a cloudburst. That tree shelters them and so does the bush. They might go into the hive or back into the bush where I could take them. But they won't fly in rainy weather, so we'll just have to wait and see."

That night the rain came and in the morning I went to see what had happened to the bees. Much to my surprise, the woman had put a section of newspaper over them to keep them dry and they were all settled in the grass, so my chance to get them in the hive was gone.

The rain continued as a light drizzle for several days, so at last I brought my hive home and left the bees to their paper shelter. Later in the week when the weather had cleared, scouts must have crept out and found a place—or the woman may have removed the paper—at any rate, the bees were gone.

I puzzled over the length of their stay and their ability to fly away, because they had been there almost a week. By all the laws of bee survival their honey supply should have been exhausted days before they left. Bees continue to surprise me and I think sometimes that the laws that entomologists set up for them need to be revised. Beekeepers can work with bees and get good cooperation, but when an unusual circumstance arises, bees will do unusual things.

An Unusual Home

In the early days of beekeeping people kept bees in all sorts of containers, pottery tubes, wooden boxes, straw skeps, and bee gums. These were all rather small, allowing the combs to hold up fairly well without the support of frames. Nowadays most beekeepers use standard hives and equipment, so I was very surprised at where a man living out near the apiary kept his bees.

My sister said, "The man who lives in that small house has a swarm of bees. He took them from one of your fence posts, but I didn't say anything because I didn't think you wanted them."

"Does he have a hive?" I asked.

"No, he put them in an old trunk with a hole cut in one end for an entrance."

"Is it a big trunk? If it is, the bees may have trouble."

"Well, that's his problem, if he wants bees and puts them in a big trunk."

I forgot all about the incident until a year later when the man moved away and I was working on material for a lecture on bees, and needed some unframed comb. I walked over to see in what condition the bees were that he had left behind. There were two trunks for bees: one was occupied and the other seemed to be empty. Both were large steamer trunks.

Slowly I raised the lid of the unoccupied one and immediately saw why there were no bees. There were no frames, of course, for frames do not come the size of a trunk, but he might have made some fairly acceptable ones that would have been a big help to the bees. The comb was built curving all around in the trunk, much like a maze, and during the heat of summer it had melted loose from the inside of the lid and lay folded upon itself on the bottom of the trunk like a thick woolen blanket. It reminded me of a fairy castle that had been toppled by an earthquake, its delicate walls all fallen.

Comb is very delicate and will crumble in one's hand if held too tightly or gouged into by an awkward thumb. There were many dead bees lying around, for the destruction must have come suddenly, and they had been engulfed in honey, much as a snowslide engulfs a person.

However, the honey had long since been removed by raiders when the trunk no longer had a protecting colony. Often there are raiding parties in the apiary when some colony is in need of honey, or is too weak to protect its stores; there are usually enough bees to put up a good fight and the result is many dead bees on the ground. There was no protection at this trunk, so the raiding must have been easy.

I removed as much of the comb as I could to show it at my lectures. The colony had been Italian, as most of ours are, the large orange-and-black-striped bees. They have nine stripes across the abdomen (four orange, five black) and the tip of the abdomen is black. The thorax is brown and fuzzy and they are beautiful bees.

The last time I checked on them the bees in the smaller trunk were still active. I made no attempt to look inside, which might have pulled the combs apart and caused another disaster.

Diseases and Enemies of Bees

Bees are subject to few diseases deserving notice; but one which may become epidemic in the apiary is foul brood, so named because it causes dying of the brood. The contagion is transmitted through the food of the larvae.

There are two types of foul brood, European and American, and both are caused by bacilli.

In American foul brood the larva is attacked when nearly ready to be sealed. It turns yellow, and gray spots appear on it. It then softens and sinks down in the cell in a shapeless brown mass. Then it becomes ropy and dries into a hard mass which the bees make no attempt to clear out of infected cells. They may even fill the cells with honey.

Sometimes the larvae do not die until capped; the cap then sinks and has a pinhole perforation in it. Such larvae dry up entirely. When the cap sinks it is much darker than the cap covering healthy larvae. It is not until the disease has made considerable progress that an odor is detected. When half the brood has rotted the odor is noticeable as a glue-pot smell, but the nose should not be relied upon, since long before the odorous stage the colony can communicate the disease to others. Only the eye can be depended on to detect the disease—and it should be a sharp, trained eye.

Several cures have been used. The fumes of salicylic acid

were used in a method devised by a Swiss apiarist, but not used extensively. Or carbolic acid in the ratio of one ounce of acid to forty pounds of syrup can be added to the food of the diseased colony. Either method requires close attention and is not safe. As long as the drugs are used the disease is held in check, but it breaks out again when the drugs are discontinued. An easier way is to use carbolic acid in the water for the entire apiary, using one ounce of acid to forty gallons of water.

Some apiarists insist that the only way to be rid of American foul brood is to burn the colonies affected with it—hives, combs, and bees. This seems a wanton waste; however, states have rules on this and they must be followed. The county bee inspector will explain.

Honey is the main source of transmission; a noted apiarist recommends removal of all combs, starving the bees for two or three days, then giving them new combs or frames and feeding them diluted honey with nutmeg and saffron added.

The safest way, the one we used, is to destroy all brood and hive honey, transfer the colony to a new hive, and give empty frames for the bees to use for several days. These frames are then destroyed and new frames with foundation given and the bees start over. This exhausts all honey that may be in the bees' system and the infection is destroyed. The hive they first occupied may be used again if it is first thoroughly scorched inside with a blowtorch.

European foul brood is not as serious as American; it is much easier to cure and less contagious, but more persistent. In this disease the brood dies a little sooner, very few larvae ever being capped, and then by perforated or shrunken caps. A small yellow spot at the head of the larva is the first sign; at death it turns black. Some of the dead larvae are removed by the bees, as there is no ropiness and they do not adhere to the cell walls. There is very little odor from the dead larvae, and no glue-pot smell.

European foul brood is cured by removing all brood and destroying it, then supplying new frames and a new Italian queen. When the diseased brood is gone, the hive is considered safe for a new queen; an Italian queen is also more resistant to foul brood than other strains.

Some beekeepers remove the queen until all brood has

hatched, but the above method seems easier. Many inspectors say it is not necessary to disinfect the hive in European foul brood.

Either type of foul brood is transmitted from hive to hive by robbing, as the bees may carry the bacillus home in the honey or on their hair. It may also be picked up from flowers contaminated by bees carrying it and show up unexpectedly in an apiary.

Not all dead brood indicate foul brood. Accidents may cause brood to die without damage to the bees, such as sudden cold weather in spring when bees have been spreading their brood or neglect or mismanagement by the apiarist in placing the brood back out of reach of the cluster after examination. Chilling can also kill brood, but it dies evenly all at once, while death from disease is scattered around in the cells.

Honey from diseased colonies is harmless to people; if it is brought to the boiling point it can even be fed to bees.

There are other, less serious diseases from which bees readily recover. Diarrhea is caused simply by the accumulation of feces in the body during a prolonged cold spell when the bees eat in the hive but are unable to fly out for comfort; bees never discharge in the hive.

"Bee paralysis" sometimes appears in apiaries and may become epidemic. The abdomen becomes distended and bees are first noticed crawling at the top of the combs as if cold and numb and looking as if their legs were paralyzed. The disease is not dangerous, but the bees lose their hair. It is a disease of the adult bee, not of the brood; there are preparations on the market for its cure.

The bee, even protected by her formidable sting, has enemies. Mice creep into hives for shelter in cold weather when the bees have retreated from the entrance and clustered on the combs, and so are unaware of intruders. The mice build a warm nest on the floor of the hive and gorge on honey, comb, brood, and bees that are too chilled to resist. They eat the head and thorax of the bees but leave the abdomen with its poison sac. A mouse will crawl in and eat until it becomes too fat to get out of the narrow entrance of the hive and dies of thirst. A colony inhabited by a mouse in winter is weak in the spring and its combs are a shambles.

Some birds are fond of bees; the kingbird devours them by

the score. He takes them on the wing (as do dragonflies), can locate hives by following the bee lanes, and is a dangerous threat to colonies. He prefers drones but also catches bees on blossoms, so workers are his prey also. It is suspected that he can distinguish between a bee in search of nectar and one returning home laden with it and in excellent condition to slide down his throat.

There is a woodpecker that pecks holes in hives and eats bees as they investigate the knocking. He can deplete a colony.

Chickens eat drones. We saw a rooster eating bees at the entrance of a hive: he carefully picked up a drone from among the bees, shook off a worker, and then swallowed the drone. Young ducks will sometimes eat bees and they may be killed by being stung while swallowing them.

Skunks eat bees as they fly out of the hive, or they scratch on the hive and eat bees and guards as they appear. Lizards do the same but the bees are unaware of them and do not attack, since the lizards are on the ground.

The toad is a well-known eater of bees. Sitting under a hive toward evening, he will sweep into his mouth, with his swiftly darting tongue, many heavily laden bees returning to the hive. It is amazing that birds and toads can swallow bees without being stung to death, but it is because they take only those heavily laden with nectar, swallowing them whole and safely storing them among other food before they recover enough to sting.

Bears are very fond of honey; in bear areas they will destroy hives unless precautions are taken. They also eat larvae in the wild; when they tear open a tree they seldom leave anything. Hives on the ground are easily turned over and the contents devoured. The apiary is heaven for bears if they are in the area; only an electric fence is an effective protection.

Ants are persistent robbers. They make nests near hives, and entering through holes or cracks, crawl up the hive and reach honey unprotected by bees. Once a few get in, more follow. Bees battle with ants, but since the ants are too small to sting, the bees bite and maul them. Injured ants are thrown out, but a mass of ants can smother a bee, pull it out of the hive, and eat it. Bees can usually keep ants in check, but a wise beekeeper will not allow an anthill to be built near a hive.

Other small enemies include wasps, hornets, spiders, and

yellow jackets. A strong colony is the best protection against small predators in the hive. Wasps rob hives, stinging bees, and biting off their heads, but a strong colony will triumph since there are many more defending bees than attacking wasps. Digger wasps catch bees on flowers, kill them, and bury them in the ground with their eggs to be eaten by the wasp larvae.

Termites may destroy wooden hives, thus endangering the colonies. There are parasites, bee lice found on the queen, that cluster around her neck and feed on the food she gets from the nurse bees. Enough lice can seriously weaken her and reduce her egg laying.

It has been estimated that a colony of fifty thousand may lose as many as a thousand bees a day to predators—about one-third of the normal population increase.

Of Moths, Mice, and Me

Another enemy of bees is the wax moth—in the adult state, a winged moth about five-eighths of an inch long with a wing spread of one-and-one-half inches. The fore wings shut together flatly on top of the back, slope steeply down the sides, and turn up at the ends like the tail of a fowl. Wax moths are gray and are seldom seen flying until dark, unless startled from their hiding places around the hive. On cloudy days just before sunset, the female may be seen trying to gain entrance to a hive, but active bees can prevent this. When the bees are at rest and she finds the door she will go in and lay her eggs in the combs.

Wax moths are very agile, both on foot and wing; the motions of a bee are slow in comparison. If disturbed in the daytime, the moths open their wings and spring or glide away. They are the bane of an apiary, because if the female is unable to get into the hive she will lay eggs in a crack on the outside and when the worms hatch they creep or burrow into the hive and do their damage. One female was seen laying her eggs in such a crack on the side of a hive. She laid about ten eggs, strung together like beads on a string, then walked about and returned to lay ten more.

Propolis is the food of the just hatched larvae, so even if the crack is sealed with it the larva has access to the hive. It is an avid eater and soon grows to adult length. It encases itself in a silken case fortified with wax so that bees are unable to sting it, and creates impenetrable galleries in a comb which it destroys by eating the wax. It prefers the comb of the brood chamber, where the cells are lined with the cast-off skins of the bee larvae, to the comb of the super.

The worms eat for three weeks, then spin their cocoons, which resemble sealed brood, in the comb cells. After ten days they hatch into moths which are already in the hive ready to lay more eggs in the combs to continue the destruction—and what is more, they live for several months in the summer.

If the worms spin their cocoons in the late fall they remain in them until spring and upon hatching have much unoccupied comb to take over. In an infected hive the worms tunnel through the combs, which become covered with a black, tangled mass of webs, cocoons, and excrement. Even a vigorous colony cannot destroy these intruders before the combs are ruined. Bees do remove larvae if possible and if they are in unoccupied comb the beekeeper can destroy them with chemicals, but many such chemicals are dangerous to use.

The worms also burrow into the walls of the hive; it is a great task to rid a colony of them after they have become entrenched. About the only way is to put the bees in a new hive with new frames, destroy the old ones, and scorch the inside of the infected hive with a blowtorch or brimstone it. Carbon disulfide may be poured on a rag in a saucer and set in the hive after covering it. One tablespoon of the liquid is enough for one hive.

Early last spring I went to the apiary to open the entrances of my three hives and was surprised to see that there were bits of wax on the bottom boards of two of them. Immediately I thought of wax moths because that is how their presence is shown. When I opened the first hive I saw with relief that the damage was not caused by them but by a mouse and all I had to do was repair the damage. The mouse was gone and I did not have to worry about anything like moth larvae being in the walls.

The mouse had crept in during cold weather when the bees were clustered on the combs and made its home in the hive. It had eaten holes larger than silver dollars in the combs, completely destroying six of them, which I had to replace. On opening the other hive I found similar damage and also the mouse. It was stiff and dehydrated, and resembled a carved wooden model more than a dead mouse, as its fur was evenly coated with wax. This colony needed six frames; as it was very small, I gave it four new frames and two of brood, plus food from the third colony which was strong and had not suffered any damage. I gave the third colony two new frames in exchange and there was no swarming that summer. I set saucers of De-Con under the hive stands and have had no further mouse problems.

The beekeeper faces an increasing problem of spray poisoning and its effect on bees, since more and more chemicals, used to control pests and weeds, also kill bees and deprive them of forage. Entire apiaries have been wiped out by indiscriminate spraying. Arsenical sprays are the worst: the bees touched are killed, and other bees drink poison from leaves and carry it home to the hive where it kills the brood. The phosphorus and chlorinated hydrocarbon groups are the most toxic to bees; malathion and other poisons have taken a heavy toll, nine-tenths of it due to careless use. Authorities in many countries are giving attention to this and legislation is in effect to safeguard bees and other innocent insects. No residue has been found in honey yet. Chemicals used should be highly specific in their action; biological control is always preferable to chemical control.

My apiary fell under the baneful influence of spraying last year. Early in the preceding summer I took a swarm from a post back of my sister's corral. It was a small swarm but taken early, and it had time to fill the hive with brood and honey. I should have checked it in the fall for strength but I was ill in the hospital and unable to do so. In the spring I found that the swarm had been winter-killed in the hive filled with honey. There had not been enough bees to make a cluster big enough to keep warm. (Ordinarily I would have combined it with one of the other colonies before winter.)

The three other colonies were thriving and I set the supers in mid-June. On visiting the apiary in July I said to my sister, "I think my bees are being poisoned."

"No one would do such a thing! What makes you think so?" she asked.

"Evidence," I said. "There are dead bees in front of the hives. Not just a few but the ground is covered with them, and it couldn't possibly be raiding to that extent."

On the first of August I checked the supers for honey harvest and to my surprise the first two contained no honey; not a single cell of the foundation had even been pulled out. I removed the supers to look in the hives, only to find all hive bees dead as well as the brood. They had died of starvation when the field bees had failed to return with nectar and the supply in the hive was used up. Sadly I took the two empty supers home and stacked the hives with some other unused ones.

The third colony had a small number of flying bees and a fair amount in the hive as well as honey, but the super was empty. I left it in place, hoping that things would improve. The colony did survive and by fall had filled four frames of the super with honey. Even a handful of bees will rebuild a colony when they have their queen; they lived through the winter and now they are my last colony.

When I found two of my colonies dead and the third one barely surviving I had the answer to what had happened. When I first moved the bees to this area it was strictly rural with fields of alfalfa, clover, and wildflowers. About four years ago a subdivision was started to the southeast and has steadily grown, converting the alfalfa fields to houses. They aren't rural homes but town houses with lawns and flower gardens, which are difficult to maintain on the sandy soil even though the lawns are sodded.

The climax was a scourge of grasshoppers which the subdivision people decided to exterminate with spray. This took most of the summer and successfully killed the grasshoppers but it also killed all the field bees. They either never returned or returned to fall dead outside the hive. There had been invasions of grasshoppers in former years but there had been no spraying because they had not seriously harmed the fields, or because the fields were then too extensive to be sprayed.

I shall not put another colony out there to meet a similar fate at the mercy of indifferent people. I doubt that they knew that bees had been killed by their spraying but I also doubt that it would have made any difference if they had known.

The spraying has just about closed the chapter of my beekeeping, but I have learned much and had a rich experience during my fifteen years of working with bees.

Bees are the most industrious and most community-oriented of all insects. No bee works for herself or her own good but for the good of the colony; she will give her life to protect it. No individual reward is asked and all that is received is food and shelter among her sisters. Man is the exact opposite: he works for himself and his to the exclusion of all others, yet man has great need of the bee to pollinate his crops and flowers. Honey, the only sweet of the Stone Age people, is still to be found on the grocery store shelves, and beeswax still has many uses.

For all man's efforts to domesticate the bee, little has changed in its way of life. Even though bees are kept in hives and tended by man they behave exactly as they originally did, often leaving their hive to settle in a tree. By providing hives, feeding when needed, and controlling swarming, man has substantially increased the productivity of bees but hasn't changed their way of life.

Index